EXISTENCE AND EXPLANATION

THE UNIVERSITY OF WESTERN ONTARIO SERIES IN PHILOSOPHY OF SCIENCE

A SERIES OF BOOKS
IN PHILOSOPHY OF SCIENCE, METHODOLOGY, EPISTEMOLOGY,
LOGIC, HISTORY OF SCIENCE, AND RELATED FIELDS

VOLUME 49

KAREL LAMBERT

EXISTENCE AND EXPLANATION

Essays presented in Honor of Karel Lambert

Edited by

WOLFGANG SPOHN

University of Bielefeld

BAS C. VAN FRAASSEN

Princeton University

and

BRIAN SKYRMS

University of California at Irvine

SPRINGER SCIENCE+BUSINESS MEDIA, B.V.

Library of Congress Cataloging-in-Publication Data

Existence and explanation : essays presented in honor of Karel Lambert / edited by Wolfgang Spohn, Bas C. van Fraassen, Brian Skyrms.
p. cm. -- (The University of Western Ontario series in philosophy of science ; v. 49)
Includes bibliographical references and index.
ISBN 978-94-010-5430-0 ISBN 978-94-011-3244-2 (eBook)
DOI 10.1007/978-94-011-3244-2
1. Science--Philosophy. 2. Philosophy, Modern. 3. Lambert, Karel, 1928- . I. Lambert, Karel, 1928- . II. Spohn, Wolfgang. III. Van Fraassen, Bas C., 1941- . IV. Skyrms, Brian. V. Series.
Q175.3.E95 1991
501--dc20 91-15399

ISBN 978-94-010-5430-0

printed on acid free paper

To Joe

TABLE OF CONTENTS

PREFACE

This collection of essays is dedicated to 'Joe' Karel Lambert. The contributors are all personally affected to Joe in some way or other, but they are definitely not the only ones. Whatever excuses there are — there are some —, the editors apologize to whomever they have neglected.

But even so the collection displays how influential Karel Lambert has been, personally and through his teaching and his writings. The display is in alphabetical order — with one exception: Bas van Fraassen, being about the earliest student of Karel Lambert, opens the collection with some reminiscences.

Naturally, one of the focal points of this volume is Lambert's logical thinking and (or: freed of) ontological thinking. Free logic is intimately connected with description theory. Bas van Fraassen gives a survey of the development of the area, and Charles Daniels points to difficulties with definite descriptions in modal contexts and stories. Peter Woodruff addresses the relation between free logic and supervaluation semantics, presenting a novel condition which recovers desirable metatheoretic properties for free logic under that semantics. Terence Parsons shows how free logic can be utilized in interpreting sentences as purporting to denote events (true ones succeed and false ones fail) and how this helps to understand natural language.

Free logic has ample historical roots and applications. Wolfgang Lenzen reconstructs the place Leibniz assigns to the notions of existence and of possible individuals within his logic which is in fact a quantificational extension of the intensional algebra of concepts. And Gordon Brittan applies Lambertian logical tools in order to shed further light on Descartes' 'cogito, ergo sum' and to construe it in effect as a plausible, if not sound argument.

Two essays jump right into old ontological questions. Peter Simons discusses whether an account of change requires objects to be continuants without temporal parts or to be temporally extended occurrents, and he defends the former view against recent criticism of David Lewis. And Kevin Mulligan is concerned with particularly problematic entities, namely with colours and orientations and their internal relations, reviewing, comparing, and in fact reviving Meinong's and Wittgenstein's opinions on that subject matter. Ermanno Bencivenga, finally, adds five nice pieces.

The other focal point of the collection is Lambert's philosophy of science. Three essays span historical space. Jules Vuillemin explains some difficulties with kinematics and dynamics and in the development of physics in general by the philosophical analysis of motion given by Plato and Aristotle. Paul Weingartner shows that Aristotle anticipated both Meinong's principle of independence, as Lambert called it, and the modern requirement of finding interpolation sentences for giving scientific explanations. And Erhard Scheibe sharply distinguishes the problems raised in the Einstein-Podolski-Rosen paradox and those raised by Bell's inequality.

Wolfgang Spohn takes up the relation between scientific explanation and understanding in the way proposed by Lambert and arrives at a positive view in which search for explanation is construed as search for coherentistic truth. It is no accident that the notion of stability is also crucial for Brian Skyrms who develops and refines a de Finettian picture of objective probability by generalizing what has been called Miller's principle and by employing ergodic theory. Daniel Hunter, finally, is rather occupied with subjective probability; he defends the method of maximum entropy updating against objections by Brian Skyrms and interprets it as a sound method of belief revision not reducible to any forms of conditionalization.

Thus the collection demonstrates the far-reaching direct and indirect philosophical impact of Karel Lambert's work for which his friends wish to offer warm thanks.

Finally three words of indebtedness: We are very grateful to Ulrike Kleemayer for preparing the index, to Kluwer Academic Publishers, in particular to Mrs. Annie Kuipers, for the effective and friendly co-operation and to Domenico Costantini who invited all three of us to a wonderful conference at the Lago Maggiore in May 1988 where the idea of this Festschrift was born.

WOLFGANG SPOHN
BAS C. VAN FRAASSEN
BRIAN SKYRMS

BAS C. VAN FRAASSEN

ON (THE X) (X = LAMBERT)

The first few years of my philosophical life were so entangled with learning from Karel Lambert that I can scarcely separate the two. In the fall of 1959 I entered my first philosophy class at the University of Alberta, a class of about seventy students with Lambert as instructor. He told us firmly that he had no intention of discussing ethics or morals, told us to read something about the pre-Socratics for next time, and dismissed the class. It became clear quite soon that he expected us to learn philosophy by doing: despite the size of the class, there was a great deal of discussion and we were constantly challenged. His response was always measured. People who were struggling found sympathy, and he would turn their questions into something significant for discussion. But students who were catching on would immediately find themselves made to face greater difficulties. It was also, I can think of no other way to put it, a lot of fun. Twice, I remember, he put me down, to everyone's amusement. The first time I wouldn't back down from a point when I was clearly losing, and Lambert ended the discussion with "Van Fraassen, you are logical, but you are not reasonable." The second time we had gotten into that subject he did not want to do at all, and I brought up Sartre's famous example of the moral dilemma of the young Frenchman, who had to choose between care for his aging mother and joining the Free French. Lambert listened patiently, then retorted: "Van Fraassen, there comes a time when a boy has to leave his mother."

The course — which like all our courses then, was a year long — ended with Russell, and the theory of definite descriptions. I was working in the university library, part time and then as a summer job, and I immediately started reading as much of Russell as I could. That was also when Lambert called me into his office, to ask me to think about going into philosophy. I said it's all right, I've already decided to do that. So he offered me a Coke, and told me to come see him and talk with him as often as I liked. Looking back on that period, and the courses I took from him, I think one thing stands out most of all. Although he was teaching us logic and related subjects, very near to his

Wolfgang Spohn et al. (eds.), Existence and Explanation, 1—18.

heart, he would never especially push or even reveal his own position. We had to come to conclusions via the problems, via questions to be struggled with, not by digesting answers. Even by the end of my fourth year, when I had become quite preoccupied with free logic, there was still the sense that it was entirely more important for me to work on problems that had become real for me, in whatever way and on whatever subject, than that I should get engaged in his.

It took in any case a while before I could even understand those. In the first summer I also started reading Reichenbach, partly because Russell had awakened my interest in the problems of space and time. Early in the second year, I read a new article by Milic Capek on eternal recurrence, in the *Journal of Philosophy.* Nietzsche was of course one of my heroes since high school, and I had thought a lot about that doctrine. Now here was an argument, based on twentieth century ideas about space, time, and relativity (my newest infatuation), and it seemed to make nonsense of this wonderful vision. I wrote a critique, applying what I had learned from Russell and Reichenbach, and showed it to Lambert. Without any meta-comments about the project, he discussed it with me, raised objections, made me clarify and rewrite — 'made me' is right, I think, for I still remember how I felt about it by the seventh and final draft — and told me to submit it. It was my first publication; but more importantly, this was how Lambert taught me in my sophomore year to do research, to criticize myself and to criticize my own criticisms, and to write. And still more importantly, to gain the sense that my ideas could be taken seriously, that ideas had the democratic right to compete, regardless of who voiced them.

Before I discuss the papers Lambert wrote in those years, let me complete this sketch of how I got to the point where I could understand them at all. In my third year, I learned about names, free logic, and the difficulties with Russell's theory of descriptions. We took up Quine's *From a Logical Point of View* and *Word and Object*, believing wholeheartedly that ontological scruples came first, and semantics a very distant second. Lambert mentioned that free logic could be interpreted by postulating some reality for non-existent objects (items treated as designata for names like 'Pegasus') but added that such an interpretation couldn't be of any interest at all. For after all, there are no non-existents, 'existence' is univocal, and the philosophical foundations of logic needed to be part of philosophy generally. I still agree with all of that, and I think he does too, and I am grateful for Quine's insistence on philosophical integrity. But in retrospect, a touch of instrumentalism would have helped. For my fourth year Lambert proposed a directed reading course, on a subject of my choice, with weekly tutorials. I opted for induction and probability, and he chose Kyburg's new book, written entirely in Quine's protosyntax, as one of the main texts. So obviously I had to study Quine's *Mathematical Logic* first, during the preceding summer. That way I also encountered limitative metatheorems in the "'yields a falsehood if appended to its own quotation' yields a falsehood if appended to its own quotation" form. At the very end of the academic year, an article appeared which showed that Kyburg had not been saved from inconsistency by this heroic attempt at formal precision. But in the meantime, I had learned a great deal more logic. I had also had the opportunity to discuss Lambert's own papers with him, with an increasing appreciation of why he wrote them in English rather than in protosyntax.

HENRY LEONARD AND THE GENESIS OF FREE LOGIC

In 1956, when Lambert was a graduate student, his teacher Henry Leonard published his seminal paper 'The Logic of Existence'. This paper set the problems in philosophical logic that preoccupied Lambert for the next ten years. With hindsight, we can discern two main problems which Leonard posed and for which he proposed solutions.

The paper begins by explaining how logic could have presuppositions, which could be removed so as to make it more widely applicable. The historical example given is the existential import of general terms in the traditional square of opposition. There the I sentence 'Some *S* is *P*' could be inferred from the corresponding A sentence 'All *S* is *P*'. Today we reject that inference, allowing only that if the A sentence is true, and there exist some *S*'s, then the corresponding I sentence is true too.

As Leonard saw it, this does not mean that traditional logic was in error. Rather it had a limitation of a sort that perhaps logic will always have, but which could be eliminated once it comes to light. The way he put it was that "traditional logic was a quite correct abstract system of logic; but . . . it was set up and developed with a tacit, or unexpressed presupposition: namely, that its terms *S*, *P*, etc., were terms having existent exemplars." (page 5)

Next Leonard pointed out that contemporary logic also had such a presupposition, with respect to singular terms. This presupposition comes to light in the inference schemes of Existential Generalization and Universal Instantiation:[1]

Sj; therefore $(Ex)Sx$
$(x)Sx$; therefore Sj

These schemes appear to yield invalid inferences if the term j is one which does not refer to anything, such as 'Santa Claus'. Again, that is not an error, but limits applicability. No errors will result if we restrict the allowable substituents for j to singular terms that have a referent. But now there is an obvious problem to address: how shall we make the presupposition explicit — i.e. state just what is allowed to replace j and what is not — and can we devise an abstract system of logic which is more widely applicable?

Leonard proposes that we characterize the category of singular terms as those terms which purport to refer to some entity, and divide them into those which do refer and those which do not. The term '*E!*' (pronounced '*E*-Shriek') which appears in Russell's theory of descriptions (though not as a primitive) he proposed for the job of marking this division: '*E!j*' is to be read as '*j* exists' and is true if and only if the term j refers to something. We are here in an area of philosophical contention, and Leonard discusses for example Quine's treatment of singular terms, and the idea (which he regarded as then prevalent) that ordinary names are all short for definite descriptions. His critique is acute; but I will restrict my exposition here to his own positive proposal.

Leonard does not shrink back from higher order quantification and modal logic. He proposes that we revise the logic of Principia Mathematica, and lays down the following principles:

(L1) $E!x$

(L2) $(Ex)Fx \equiv (Ex)(E!x \,\&\, Fx)$

(L3) $(x)Fx \equiv (x)(E!x \supset Fx)$

(L4) $E!(Ix)Fx \equiv (Ey)(x)(Fx \equiv x = y)$

(L6) $Fy \,\&\, E!y \cdot \supset \cdot (Ex)Fx$

(L7) $(x)Fx \,\&\, E!y \cdot \supset \cdot Fy$

(L8) $(x)Fx \supset (Ex)Fx$

I will leave L4, a principle concerning definite descriptions, for discussion below. Omitted altogether is L5, which pertains to a separate proposal concerning the possibility of defining *E!* by means of higher order quantification into modal contexts. On page 60, before stating the above principles, Leonard proposed to take *E!* as primitive for the time being, and I shall just stick to that here. Finally, Leonard himself points out that L1 and L8 together entail $(Ex)E!x$, i.e. that at least one thing exists. A little later various authors, including Lambert, saw that as a still further presupposition which limits the applicability of logic as well.

What is not clear from Leonard's paper, is just how much the above principles are meant to do. To what are they to be added, so as to yield a satisfactory system of (first order) quantificational logic? That is a question about completeness, and Leonard was not in a position at that point to give that question a precise content. The history of this and related problems is treated adequately in the introduction and selections included in Bencivenga's *Le Logiche Libere.* Roughly speaking, satisfactory systems of that sort came into the literature in 1959, at the hands of Hailperin and Leblanc, and Hintikka. In his abstract of a paper for the International Congress of Logic, Methodology, and Philosophy of Science at Stanford in 1960, Lambert introduced the term 'free logic' to stand generally for systems free of presuppositions of the sort Leonard had discussed.

That was the first problem Leonard set, and his partial solution. Let us take a careful look at its general character. Quite in accordance with his way of introducting the problem, Leonard thought of free logic (as I shall now continue to call it) as a fragment of the standard logic. The theorems of free logic as he sketched it, were part of the theorems of *Principia Mathematica*. But the class of terms that qualified as substituents for free variables in the axioms and rules, was larger, it was not restricted to referring terms. Bound variables were unaffected, for the quantifier retained its standard interpretation. As Leonard codified it:

> We agree with [Quine] that "To exist is to be the value of a variable." But our revised logic is such that we disallow his claim that to name an existent is to be a substituent of a variable. Instead, our logic comports with "To purport to name an existent is to be a substituent of a variable." In other words, not all substituents designate values. (page 60)

When I came to Pittsburgh as a graduate student in the fall of 1963, I took a seminar from Nuel Belnap, with as main topic the logic of questions. It began with an introduction to the semantic analysis of logic, and Belnap presented the short, elegant completeness proof for quantificational logic that he and Alan Anderson had recently published. In a review, Belnap also indicated how this can be amended for free logic, with the singular terms all being given referents in a domain only part of which supplies the range for the quantifier. The logic of 'Santa Claus' is then regarded through the fiction that Santa Claus does exist, but outside the class spanned by our 'All' and 'Some'.

Looking through my correspondence with Lambert during that year, I found a letter from Lambert, responding to what must have been one from me about what I had been learning there.

Yes, I was quite well aware of the sort of proof of completeness Belknap (*sic*) suggested to you for "free" logic — as much as four years ago. My proof was parallel to his; split the domain into real and imaginary objects, restrict the range of the quantifiers to real objects, replace Specification by Ramified Spec., and proceed a la Henkin. Hintikka suggested this sort of proof to me, though I already knew how to do it. But like you, I have been trying to find a "standard" model, as you call it. (Nov. 26, 1963)

Had the problem — the first problem Leonard set — been solved satisfactorily? In my mind, and as the last sentence of this passage indicates, in Lambert's mind, the answer was *no*. It had been shown that the indicated fragment of standard logic had a certain autonomy: it was exactly the logic obtained from the standard one by restricting the quantifier to the extension of a non-empty but otherwise arbitrarily chosen predicate. Perhaps the right understanding of 'exists' entails that its logic must coincide with that logic — or perhaps there is more to it.

But the problem had been handled well enough to give a satisfactory setting for dealing with the second main problem which Leonard had set — and which preoccupied Lambert a good deal more.

LEONARD ON DEFINITE DESCRIPTIONS

This second problem was to develop an adequate free logic of descriptions. Leonard rejected Russell's theory, at least partly for the same reason as before: that the logical treatment of terms should be uniform, regardless of matters of fact such as whether this or that thing exists. The above departure with respect to terms generally allows a new treatment of descriptions too. The first 'big change', says Leonard, is that definite descriptions are allowed as substituents for free variables in the theorems of logic. As first example, Leonard gives

(L11) $(Ix)Fx = (Ix)Fx$

as substitution instance of the law of self-identity. This is not a theorem of *Principia*; but on the other hand, some of those theorems will have to be rejected. Recall that Leonard had already listed for retention

(L4) $E!(Ix)Fx \equiv (Ey)(x)(Fy \equiv x = y)$

which was Principia *14.02. He also retained half of Principia 14.01, namely

(L14) $(Ey)[(x)(Fx \equiv x = y) \,\&\, Gy] \supset G(Ix)Fx$

But the other half, the converse of the L14 would via L4 entail

(a) $G(Ix)Fx \supset E!(Ix)Fx$

Could that be maintained? Not for arbitrary open formulas G . . . So at best we would land into the mess of scope problems, trying for example to retain (a) for primitive predicates G, but distinguishing between $[-G](Ix)Fx$ and $-[G(Ix)Fx]$. After all, the new policy for substitution generates as theorem

(b) $H(Ix)Fx \vee -H(Ix)Fx$

so that H . . . and $-H$ . . . could not both replace G . . . in (a) without yielding the disastrous theorem

(c) $E!(Ix)Fx$

In another respect, however, Leonard wanted to go quite definitely beyond *Principia*, which has

(*14.22) $E!(Ix)Fx \equiv F(Ix)Fx$

Leonard comments "The right-hand member appears to be analytic, and should be assertible without restriction to descriptions that exist." I cannot make out what the last two words are doing here, but it seems that he wanted something stronger than *14.22. Leonard's next statement certainly denies, however, that the attribution "appears to be analytic" can be taken at face value. In the case of a predicate F for which ($Fx \supset E!x$) is analytic, he says,

(d) $F(Ix)Fx$

cannot be generally true. Specifically, if $F = E!$, then (d) would be

(e) $E!(Ix)E!x$

which says, via L4, that only one thing exists.

Leonard goes on to propose an extension of the little theory of descriptions sketched so far (L4, L11, L14) within modal logic. The problem which came to concern Lambert, however, was the theory of descriptions for first order logic with identity alone. This is what I had in mind as the second main problem Leonard set for his philosophical posterity.

Because the theory of descriptions will be our focus, I will now provide the important formulas with mnemonic manes.

(LEON-EX) $E!(Ix)Fx \equiv (Ey)(x)(Fy \equiv x = y)$

(LEON-ID) $(Ix)Fx = (Ix)Fx$

(LEON-AT) $(Ey)[(x)(Fx \equiv x = y) \,\&\, Gy] \supset G(Ix)Fx$

(MEINONG) $F(Ix)Fx$

Here EX, ID, AT stand for *existence, identity, attribution*, and MEINONG is mean to insinuate something about Meinong — in fact the latter's principle studied by Lambert later on, of the independence of Sein and Sosein.

LAMBERT'S EARLY ESSAYS ON DESCRIPTION THEORY

In 1959 Jaakko Hintikka had published two papers, one each for the two problems set out above. The second, 'Towards a theory of definite descriptions' appeared to provide just what was called for, a theory of definite descriptions, allowed as substituents for free variables in the theorems of first-order free logic with identity. But Lambert, in his 'Notes on *E!* III', pointed out that (MEINONG) above is provable in Hintikka's system, which has axiom

(HINT) $y = (Ix)Fx \equiv \cdot\, (x)(Fx \supset x = y) \,\&\, Fy$

Via (LEON-ID) this leads to (MEINONG) and also — as Lambert points out further on (page 56) — to

$$(x)(Fx \supset x = (Ix)Fx)$$

which has equally disastrous consequences.

Hintikka replied in 1964, proposing to repair his system by rejecting (LEON-ID). To do this, he had to reject substitution of definite descriptions for the free variable in his general axiom ($y = y$) of self-identity. The paper notes that the restriction must not be easily circumventable, but suggests still that it is a matter only of restricting such substitution. Lambert pointed out in a later rejoinder (1966) that the restriction would have to be extensive — perhaps universal. For suppose we allow the inference from (HINT) to

(g) $y = (Ix)(x = y) \equiv (x)(x = y \supset x = y) \& y = y$

where $F \ldots$ is replaced by $\ldots = y$. Then we can infer to

(LAMB-ID) $y = (Ix)(x = y)$

If now the restriction does not forbid replacement of y by $(Iz)Fz$, that gives

(h) $(Iz)Fz = (Ix)(x = (Iz)Fz)$

Now, again if the restriction allows it, substitute $(Iz)Fz$ for y in (g) to get, via (h), to

(i) $(x)(x = (Iz)Fz \supset \cdot x = (Iz)Fz) \& (Iz)Fz = (Iz)Fz$

which yields (LEON-ID) after all. At this point it is not clear just how much Hintikka's restriction should forbid; the problem seems to be out of hand.

In 'Notes on *E!* III' Lambert had advanced his own proposal. He proposed (in the context of a general rule that all singular terms — both names and descriptions — are substitutable for free variables in the general theorems of logic) the special principles:

(LAMB-ID) $y = (Ix)(x = y)$

(MFD) $(y)[y = (Ix)Fx \equiv \cdot (x)(Fx \supset x = y) \& Fy]$

The tag MFD was introduced later by Lambert as mnemonic for 'Minimal Free Description theory' (see his 1972). Note that it restricts Hinitikka's basic principle to existents — it says under what conditions an existent is $(Ix)Fx$. For non-existents the information we get is only that from general logic — including therefore (LEON-ID) — plus the special principle (LAMB-ID). But in the meanwhile, existence had been proved to have its natural general explanation in free logic, namely

$$E!y \equiv (Ex)(x = y)$$

and via this biconditional, we get at once from (MFD) to (LEON-EX) and (LEON-AT).

In his 'Notes on '*E!*' IV', Lambert proposed a strengthening of this theory. The new single principle to be added to first-order free logic with identity was

$$\text{(LAMB-IV)}\quad y = (Ix)Fx \equiv (z)[z = y \equiv \cdot\, Fz \,\&\, (x)(Fx \supset x = z)]$$

The tag IV just refers to the title 'Notes on '*E!*' IV' of course. Note the difference from (HINT): this principle gives the truth conditions for ($y = (Ix)Fx$), and it agrees with (HINT) when the terms y and $(Ix)Fx$ refer. But if one refers and the other doesn't, it makes the statement false — as Leibniz's law of course requires — while if neither term refers, it makes the statement automatically true.

This is a stronger theory, for it implies both (LAMB-ID) and (MFD). In addition, it implies, as we just saw, in effect:

$$\text{(LAMB-NONEX)}\quad -E!y \,\&\, -E!(Ix)Fx \cdot \supset y = (Ix)Fx.$$

That is all; (LAMB-IV) is in turn entailed by (MFD) and (LAMB-NONEX) together.

THE SPECTRUM OF FREE DESCRIPTION THEORIES

Leonard had divided the logic of descriptions into two parts for free logic. The first part, encapsulated by Lambert in (MFD) says in effect that Russell was right in his treatment of those definite descriptions which do refer. The second part was the set of principles to be added to free logic, elaborating on the meaning of 'the so and so' generally, and applicable as well to those definite descriptions which do not refer. While the first part is a complete and unique solution to its proper problem, the second part allows for a whole spectrum of options, from complete neutrality (accept (MFD) alone), to a complete and arbitrary fiat for the treatment of all non-referring terms, such as (LAMB-IV).

Should we think that there must be a uniquely right point on this spectrum? Even if the answer were *yes*, we might have to add that the understanding of 'the so and so' cannot be complete within the compass of first-order free logic with identity. Leonard was perhaps right to think that this understanding would also require a study of 'the' in modal discourse — at least. A unique right point on 'our' spectrum (of first order theories) would then be entailed by the right account in that richer logical context.

But I suspect that in the richer context, all we will find is a more richly nuanced spectrum. For there is a general problem that arrives with the modalities: exactly that we can start cataloging alternative possibilities. Let us, like Leonard, ask how (MEINONG) might be restricted so as to yield a valid general principle. Leonard proposed

(LEON-MOD) $-(x) \Box (Fx \supset E!x) \supset \Box F(Ix)Fx$

But now consider

$Fx =$ [x is such that exactly N objects exist].

This can be asserted of x without entailing that x exists. So by (LEON-MOD) we get

(Ix)(x is such that exactly N objects exist)
is such that exactly N objects exist

which is false for most N. Clearly, (LEON-MOD) is not saved by adding to its antecedent

$\ldots \& \Diamond Fy$

for after all, you or I could be such — i.e. inhabit a world such — that there are exactly N things, for many numbers N.

The only restriction that would prevent this sort of problem is one that guarantees that Fy does not conflict with any of the facts about existents in our world. But what if F says nothing about the real things? Let us assume the following as given:

$-E!y \& Fy$

It is clear now that $(Ix)(x = y \& -Fx)$ does not exist. It also can't be a non-existent, so to say, or at least, we can't assert

$(Ix)(x = y \& -Fx)$ has the property that it is identical with y and is not F

though I don't see what modal fact could give the right restriction for (LEON-MOD) to prevent this as consequence. After all, $(Ix)(x = y \& -Fx)$ possibly exists, is possibly identical with y, and so on. So if we want to restrict (MEINONG), the predicate should not even be such that, if it applies truly anywhere, that could contradict given facts about anything, existents or non-existents. That leaves us only with necessary properties. But they should be very necessary. That is, if *F* is to qualify for use in (MEINONG) so as to yield a truth, we would need a guarantee that there is no term *t* at all such that *Ft* could contradict anything else. And that we can't express unless we add a new sort of quantifier – call it $(/x)$ – such that Universal Instantiation does hold for arbitrary singular terms when it comes to that quantifier.

Perhaps we could, meaningfully, have that. Lambert and I explored this sort of quantifier in a later joint paper. But at this point in our present argument, such principles don't give us any useful information. We are near to the stipulation that *Fy* itself must be a theorem – in which case the relevant instance of (MEINONG) is a theorem already – and are not genuinely told any more. The answer to the question about what is true about non-existents is still left entirely undetermined.

QUESTIONS OF COMPLETENESS

Just as Lambert's main problems were initially set by his teacher, so Lambert set my initial problems in philosophical logic. The first derived from his insistence that a philosophically illuminating semantics for free logic would have no recourse to entities outside the range of the quantifiers, as surrogate designata for non-referring terms. In the fall of 1963, shortly after I had come to Pittsburgh as a graduate student, I wrote Lambert about this. Whatever proposal I made, I doubt that it had much to it; he answered

> Later, I shall comment on your proposal for proving the completeness of free logic. You are right. Your proposal differs from Belknap (*sic*) and mine only in a theoretical way. But I have some misgivings, nevertheless, about the way you propose to establish completeness via the technique of models. (Dec. 18, 1963)

In the summer of 1964 I wrote a paper for a directed reading course with Nicholas Rescher on tense logic, 'Tense logic for corruptible entities', in which the quantifiers in present tensed sentences (devoid of tense-modal operators) were interpreted as ranging over presently existing objects only. Since the language contained names that could refer to people already dead or not yet born, that meant that the logic of the present tense fragment had to be free. With that excuse, the paper became the vehicle for work in free logic, and I gave the semantics and completeness proof (which I wrote up properly later on in 'The completeness of free logic'). On August 28, 1964 I sent relevant parts of the essay to Lambert, together with an exposition called 'Two interpretations of free logic' in which I explained why and how free logic should and could be neutral between the various proposals about assigning truth values and leaving truth value gaps when it comes to sentences like 'Pegasus has a white hind leg'. On September 9 already Lambert sent me a letter about it, and three pages of comments, to which I replied two days later — well, when you're hot, you're hot!

Through Lambert's wholehearted and charitable interest in what I was doing, correspondence and conversations almost imperceptibly grew into joint work.[2] In February 1965 we were corresponding intensely about Lambert's 'Notes on E!' (III and IV), and had in effect begun work on our eventual joint paper 'On free description theory'.

Let me explain briefly about how the formal context had changed. Since Lambert had made me learn the system of Quine's *Mathematical Logic*, that is what I began with. Now this is a system designed for a language in which there are no names, and in which formulas containing free variables are not genuine statements at all — just formal conveniences. If we now just add names to this language, and keep everything else the way Quine had it, we have a free logic already. For the formula '$x = a$' takes its place among the formulas containing a free variable, and is treated no differently from 'Fa', where 'F' is an arbitrarily chosen monadic predicate. Of course this is uninteresting until we add principles that have to do with names, specifically:

$$\vdash t = t$$
$$\vdash t = t' \supset \cdot A \supset A(t'/t)$$

where Quine's $\vdash$ indicates that the universal closures of what follows are all theorems (and where t, t' stand for any singular terms or free variables, and (t'/t) is the substitution of occurrences of t' for all occurrences of t).

Understood in this way, free logic is developed not as a fragment of standard logic, but as an extension; yet it is obviously the same thing — in so far as quantificationally closed formulas are concerned — as what we had before. Specifically, we can prove e.g.:

Suppose Fa and $(Ey)(y = a)$. Then $(Ey)(Fa \mathbin{\&} y = a)$.
But $\vdash Fa \mathbin{\&} y = a \cdot \supset \cdot Fy$; therefore, $(Ey)Fy$.

which is the right, free version of Existential Generalization.

Suppose we forget about names, do not add them, but add the description operator I to the syntax, and let the singular terms (substituents for t, t') be just exactly the free variables and the expressions $(Ix)A$, where A is any well-formed formula. Then what goes for 'a' above will obviously go for '$(Ix)A$'. What could be added by way of axioms to yield a free description theory?

The identity axioms above already tell us implicitly that $(Ix)Fx$ can be identical with at most one thing; any additional principle which applies to referring descriptions serves therefore just to narrow down which existent $(Ix)Fx$ is, if it exists.

The question is now what we should add as principles for description theory, to cover also the non-referring cases. If like Leonard we agree that Russell was right in the case of referring descriptions, we still have to worry about his bothersome scope rules. But if Lambert's discussions taught us anything, it is that it will suffice to concentrate on simple assertions of identity. So the agreement with Russell should be expressible as:

(RUSSELL-ID) $\vdash E!(Ix)Fx \cdot \supset \cdot (y)(y = (Ix)Fx \ \cdot \equiv \cdot Fy \ \& \ (x)[Fx \supset x = y])$

But of course, since the quantifier ranges only over existents, tautologically so, we also have:

$$\vdash -E!(Ix)Fx \cdot \supset \cdot (y)(-[y = (Ix)Fx])$$

But Russell had said, and Leonard agreed to this too, that if there is exactly one F, then $(Ix)Fx$ exists:

(RUSSELL-EX) $\vdash (Ey)(Fy \ \& \ (x)[Fx \supset x = y]) \supset E!(Ix)Fx$

It was Lambert's insight that putting all this together amounts exactly to the main principle of his 'Notes on *E!* III', namely (MFD).

Thus we have a rock bottom minimum for free description theory: the one that says when $(Ix)Fx$ exists, and says which existent it is identical with if it does exist. This closes the book, so to say, on the treatment of referring descriptions. There is nothing that can be subtracted without leaving a glaring open question. And there is nothing that can be added which pertains just to the referring ones, for there is nothing — describable in the object language — that is common to just those entities which are designated by definite descriptions. So, any additions will have to be about the non-referring cases.

Here we have just one sort of guidance from the previous literature: the persistent temptation, leading into error on a number of occasions, to subscribe to

(MEINONG) $\vdash F(Ix)Fx$

So one sort of addition will be to assert (MEINONG) for a restricted, 'safe', class of cases. That is just the sort of addition Lambert made in 'Notes on E! III', with

(LAMB-ID) $\vdash (Ix)(x = t) = t$

here written so as to show how it is a special case of (MEINONG). Rich Thomason had another candidate at one point:

(THOMASON) $\vdash Ft \,\&\, -E!t \cdot \supset \cdot F(Ix)Fx$

i.e. $(Ix)Fx$ is F if anything unreal is. My own single contribution to the eventual 'spectrum of free description theories' was of a different sort, having nothing to do with (MEINONG) directly. I proposed a rule which, like Universal Generalization, generates theorems from theorems, rather than conclusions from premises *ueberhaupt*:

(vF) if $\vdash Ft \equiv Gt$ then $\vdash (Ix)Fx = (Ix)Gx$

The mere fact that this rule is not derivable from the foregoing had disturbed me — but of course, if you are not saying anything about non-existents, and you know that $(Ix)(Fx \,\&\, Gx)$ does not exist, then you do know that $(Ix)(Gx \,\&\, Fx)$ does not exist either, but not whether they are identical or distinct!

A strengthening of (vF) can be suggested, but only if we do not care too much about non-existents — if we don't think that there are interesting true stories about Pegasus and the like. In that case, we might be willing to accept:

$$(x)(Fx \equiv Gx) \supset (Ix)Fx = (Ix)Gx$$

with its corollary that the existent golden mountain *is* the golden mountain, and that if in fact no real men eat quiche, then the man = the man who does not eat quiche. I was not willing to go so far. Lambert, in 'Notes on E! IV' had been willing to go further, as I pointed out above. The last two additions I have discussed saw the light later, but in our joint paper, we exhibited the system of 'Notes on E! IV' as the joining of Russellian and Fregean approaches to description theory. For as we showed, it can be produced by adding to the minimal free description theory either, the *E!* IV principle suitably formulated:

$$\vdash t = (Ix)Fx \cdot \equiv \cdot (y)(t = y \equiv \cdot Fy \,\&\, (x)[Fx \supset x = y])$$

(where x and y are distinct variables), or the principle that any two non-existents are identical:

$$\vdash -E!t \,\&\, -E!t' \cdot \supset \cdot t = t'$$

or else the triple of principles, two marking agreement with Russell, and the third the Fregean principle which Kalish and Montague had introduced in their treatment of descriptions:

$$\vdash E!(Ix)Fx \equiv (Ey)(x)(Fx \equiv x = y)$$
$$\vdash (Ey)(x)(Fx \equiv x = y) \supset F(Ix)Fx$$
$$\vdash -(Ey)(x)(Fx \equiv x = y) \supset (Ix)Fx = (Ix)(-[x = x])$$

Short of adding a specific story about the non-existent, this is the strongest free description theory one can have — it reduces the possible stories that can be true about non-existents to those which have only a single character. At the conclusion of the article we suggested that this might be appropriate in special, technical contexts, such as the foundations of mathematics, but not in general.

CONCLUSION

When I set out to reconstruct what happened at the conception, birth, and maturing of free logic and description theory, I decided to limit the story to what I had seen from close up at the time. It brought back a wealth of memories of how Karel Lambert had allowed me to enter and participate in this intellectual journey — beginning already well before I understood very well what was going on. His example as a teacher, which I have tried to keep as a model, was never to proselytize or dictate, but to provide rich and varied opportunities for learning. His encouragement consisted especially in giving us the feeling that any ideas we came up with would be treated as possible contributions, as quite possibly containing great promise. Then he would point us to new literature — quite often, and without saying so, including views at odds with his own — and suggest new problems, again as opportunities we could take to advance our thinking. Then if we did come to learn his views, they would appear as solutions to problems which we had struggled with ourselves, and after we had gained some independent ability to evaluate the proffered solutions.

Reading this last paragraph over again, I realize that in trying for a summary of my debt, I have produced something rather abstract, and perhaps a little stilted. I remember quite well Lambert's ironic smile when I'd go off into something pretentious, pedantic, or overly abstract. I have tried to keep that in mind as well, even if not always equally successfully. But I wanted to explain just how much I learned, much of it not at all theoretical, in learning how to answer the question *who* = $(Ix)(x = \text{Lambert})$?

Philosophy Department,
Princeton University
Princeton, U.S.A.

NOTES

[1] The use/mention distinction will not be strictly observed, but in general my symbols are not object language expressions but names thereof, or placeholders for such names.

[2] Lambert had moved from Alberta in the summer of 1963 to become chairman of the Philosophy Department of West Virginia University, in Morgantown, W. Va. Since that is not very far from Pittsburgh, it was possible to meet from time to time. His letters make frequent mention about his trips to Pittsburgh to take part in golf tournaments, and our meetings on those occasions. Eventually Lambert hired me as a full time instructor in his department for the spring term of 1966, while I was finishing my dissertation at Pittsburgh.

REFERENCES

* indicates inclusion in Bencivenga

Bencivenga, E. (ed.): 1976, *Le Logiche Libere*, Boringhieri, Torino.
Hintikka, J. J. K.: 1959, 'Towards a theory of definite descriptions', *Analysis* **19**, 79–85.
Lambert, J. (Karel): 1958, 'Note on 'E!'', *Philosophical Studies* **9**, 60–63.
Lambert, J. (Karel): 1959, 'Singular terms and truth', *Philosophical Studies* **10**, 1–5.
Lambert, J. (Karel): 1961, 'Notes on 'E!': II', *Philosophical Studies* **12**, 1–5.
*Lambert, J. (Karel): 1962, 'Notes on 'E!': III, a theory of descriptions', *Philosophical Studies* **13**, 51–59.
Lambert, J. (Karel): 1963, 'Existential import revisited', *Notre Dame Journal of Formal Logic* **4**, 288–292.

Lambert, J. (Karel): 1963, 'Some remarks on singular terms: a critical review of H. S. Leonard, *Principles of Right Reason*', *Inquiry* **6**, 257—272.

Lambert, J. (Karel): 1963, 'Quantification and existence', *Inquiry* **6**, 319—324.

*Lambert, J. (Karel): 1964, 'Notes on 'E!'': IV, a reduction in free quantification theory with identity and definite descriptions', *Philosophical Studies* **15**, 85—87.

Lambert, J. (Karel): 1965, 'Logic and existence', *Notre Dame Journal of Formal Logic* **6**, 135—141.

Lambert, J. (Karel): 'Definite descriptions and self-identity II', *Philosophical Studies* **17**, 35—43.

*Lambert, J. (Karel): 'Free logic and the concept of existence', *Notre Dame Journal of Formal Logic* **8**, 133—144.

Lambert, J. (Karel): 1972, 'Notes on free description theory', *Journal of Philosophical Logic* **1**, 184—191.

*Leonard, H. S.: 1956, 'The logic of existence', *Philosophical Studies* **7**, 49—64.

Van Fraassen, B. C.: 1966, 'The completeness of free logic', *Zeitschrift für Mathematische Logik and Grundlagen der Mathematik* **12**, 219—234.

*Van Fraassen, B. C. and Lambert, K.: 1967, 'On free description theory', *Zeitschrift für Mathematische Logik und Grundlagen der Mathematik* **13**, 225—240.

ERMANNO BENCIVENGA

FIVE EASY PIECES

Lambert has shown in a number of occasions how to make significant philosophical points while using space very economically. The following pieces are conceived as homage to his example in this regard.

1. FREGE'S OTHER DUALISM

One way of understanding Frege's distinction between concept and object is as an approach to the problem of the third man argument: if concepts are irreducible to objects, to the point of not even being possible objects of discourse, then that argument cannot get started. Here I will show that another Fregean distinction — that between grasping and asserting a thought — can be seen as an approach to another Platonic puzzle: the reduction of knowledge to a form of recollection. I leave it to others to decide whether this analogy points to a more general, deeper opposition between the conceptual structure of Frege's realism and that of other, more traditional variants of the same general position.

According to Frege, there would be no objectivity to human knowledge if it weren't for thoughts. It is by mobilizing thoughts that one can account for the fact that what I know when I know the Pythagorean theorem is the same as what you know when you know that theorem. The problem is: how exactly is this 'mobilization' supposed to work?

Thoughts are expressed by sentences, they are the senses of sentences, and understanding a sentence means grasping the thought expressed by it. So suppose that grasping a thought be the only relation one can have with it. Suppose also that I am sitting in a windowless, soundproof room, and you come in and tell me:

(1) It is raining outside.

If (1) is true and I have reasons to believe you, we will consider this a case in which knowledge of (1) is transmitted from you to me. But, even before your utterance, I was in a position to grasp the sense of (1), that is, more specifically, I grasped the senses of the component expressions

Wolfgang Spohn et al. (eds.), Existence and Explanation, 19—24.

'raining', 'outside', etc. and I knew how to combine these component senses into the sense of a sentence. In fact, transmission of knowledge *could not* have occurred in this case if I had not been in a position to grasp the sense of what you said. And here is where the problem comes up. For suppose that I analyze the fact that I have come to know (1) as a result of your uttering it in terms of my grasping the sense of (1); then your utterance would be totally superfluous, and I could have reached the same 'knowledge' by articulating my linguistic competence in total isolation.

In conclusion, if thoughts are to be involved in (the acquisition of) knowledge, the relation I have with them when I (come to) know something must be different from, and irreducible to, the relation I have with them when I understand the sentences expressing them.

2. TRUTH CONDITIONS

Ever since material implication (MI) came onto the scene, there have been claims that it could not be used to adequately represent conditionals, and proposals to replace it with stricter, stronger, more 'relevant' relations. But consider the following justification for the truth conditions of MI. Suppose I say today

(2) If it rains tomorrow, I will take my umbrella.

and ask yourself under what circumstances tomorrow you could call me a liar. Clearly, only if it rains and I did not take my umbrella. Only in that case, then, would (2) be false. If we accept bivalence, in all other cases (2) would be true.

If you find this justification convincing, as most of my students do, you may wonder what did the trick. The answer is: you (implicitly) defined 'true' as 'what is not false'. If you had done the converse, you would have no clear way to go if tomorrow it does not rain.

Attributing conceptual priority to falsity over truth has vast practical and philosophical implications. It makes truth into a public matter, something that withstands objections and criticism, rather than something that one can convince oneself of through the private experience of being struck by evidence. It invites to make ambitious claims, even when we are not completely clear about them, since (who knows?) we might be able to respond to all attacks *actually* made on them. And in the presence of uncertainty, which is our ordinary epistemic state, it makes for more truth than falsity, it legitimizes more conflicting voices so long as none of them can be conclusively shut up by the opposition.

Given all these implications, it is of interest that the conceptual priority of falsity be found at the very heart of our standard logic: in that MI through which the crucial logical notion of inference is often defined. This fact also gives an interesting perspective on those who repeatedly rebelled against MI, in the name (we can say now) of a more conservative notion of truth.

3. SET THEORY WITH CREATIVE DEFINITIONS

The requirement that mathematical definitions be non-creative was part and parcel of some ambitious programs in the *philosophy* of mathematics: specifically, logicism and formalism. And it made good sense within those programs. If you want to establish that all mathematics is logic (that is, that all mathematical theorems are logical truths), you will want to avoid the possibility that new theorems be 'sneaked in' by defining new symbols. And if you intend to prove the consistency of a mathematical theory, you will not want to worry about the fact that (assuming you have once completed your task) a new definition might reopen the can of worms.

But now those philosophical programs have failed, and the issue of whether creative definitions are to be allowed can be taken up again, evaluating the pros and cons in the *new* situation. Among the cons, clearly, there is the fact that creative definitions force one to face a (possibly) new theory (if a theory is a set of theorems) every time a new definition is given. They force one to a less ultimately foundational, more case-by-case approach to theories: every time one feels the need for a new symbol, one may have to reconsider the theory as a whole. Here I want to point to a pro for a specific kind of creative definition.

Suppose you can define a symbol in context, and then use it outside that context, on the basis of whatever its grammatical category is. For example, you define the power set $P(S)$ of a set S by

$$(3) \qquad x \in P(S) =_{df} \forall y(y \in x \supset y \in S)$$

and then use $P(S)$ (not necessarily in the context $x \in P(S)$!) as a substituend for individual variables. This strategy clearly makes for creative definitions; in fact, if you used it to define the Russellian set R

by

(4) $x \in R =_{df} x \notin x$

you could turn a provably consistent set theory into an inconsistent one. But that is only to stress the fact that creative definitions would require a more individualized, less wholesale approach to the practicability of conceptual tools. An *advantage* of the strategy, on the other hand, is that it would make the surface grammar of 'naive' set theory much closer to the deep grammar of axiomatic set theory. In naive set theory, $P(S)$, R, and the like *are* singular terms, and if one is asked 'what one means' by them, one is likely to answer by some variant of (3), (4), and the like. In axiomatic set theory, however, such terms are usually introduced by circuitous existential axioms that make (the logical form of) proofs quite convoluted (see also below). That, of course, would be a price one is happy to pay if the deep grammar displayed by axiomatic set theories paid some philosophical dividends, but, after the demise of logicism and formalism, I don't see what can be invoked here as a payoff if not some notion of conceptual 'economy'. And, if that notion is invoked, I find it useful to note that it should be balanced against considerations of naturalness like the ones brought out here.

(Of course, it would be possible to consider (3) and the like as *axioms* rather than definitions, but this would entail that (not only the theory, but also) the (primitive) *language* of the theory might have to change every time we add some such 'axiom' or, alternatively, that we would have to decide in advance all the *possible* symbols we will ever need — an assumption of omniscience that would fit the philosophical programs mentioned above, but is hardly palatable in light of their failure.)

4. IMPERFECT PREDICATION

It has been claimed that there exists a tendency, in some people, to identify things sharing a common predicate. This 'thought disorder' has even been considered by some authors (originally, Von Domarus in a paper of 1944) as a law of 'schizophrenic logic'. Here I want to point out that the disorder could be seen as a consequence of an imperfect performance of the operation of predication.

Suppose you have a language L whose vocabulary is the same as that of monadic predicate logic with identity and descriptions. The formation rules of *L*, however, are quite peculiar, in that the only way a predicate *P* can be used in it is by constructing the description 'the *P*', and that is also the only way descriptions can be constructed. The only atomic sentences of *L*, therefore, are identities.

In *L* you cannot construct anything of the form

(5) The *P* is *Q*

(for example, 'The table is brown'). How then can you best approach the information content of (5) in *L*? It seems you would have to say things like

(6) The *P* is (the same as) the *Q*

(for example, 'The table is (the same as) *the* brown (thing)'). But if that is the only way you can say it, you will be forced to 'inferences' like

(7) (a) The *P* is (*the*) *Q*
(b) The *R* is (*the*) *Q*
(c) Therefore, the *P* is (the same as) the *R*.

(A crucial feature of *L* is that it does not adopt the usual strategy of defining descriptions *in terms of* predication (for example, defining 'the *P*' as 'the *x* such that *x* is *P*'). If that were the case, one could use sentential operators to construct *complex* descriptions, and then paraphrase 'The *P* is *Q*' more adequately as 'The *P* is the (*P* & *Q*)'.)

5. CAN GOD NOT BE A CREATURE?

The question whether God created himself is somewhat analogous to the question whether a set is a part of itself. We know that the answer to the latter is: Yes, it is, an *improper* part. Which suggests that there is something 'improper' to calling a set a part of itself; however, the way such 'impropriety' is ordinarily characterized is by *fiat*, that is, by definition (a proper part of *S* is a part of *S* that is distinct from *S*). We could do the same with God, of course, and make it part of the definition of '(proper) creature' that God not be one. The problem I want to consider is: is there any more *interesting* way of ruling out that possibility?

One difficulty here is that many philosophically sophisticated accounts of creation do not conceive of it as a temporal event. God is not in time: he creates time as much as he creates anything else, and there is no reason in sight why he could not also create eternal beings like himself.

What does then his 'creation' amount to? One popular answer is that it consists of providing a logical and/or metaphysical ground for whatever is created. But clearly God is a logical and/or metaphysical ground for himself. Or one might want to add will to the matter, and say that God willed something to be. But again, it is commonly believed that God wills himself, indeed necessarily so.

Some notion of ontological dependence comes naturally to mind: a creature *a* is dependent on God in the sense that, were God not to exist, *a* would not exist either. But even this won't work: if God were not to exist, trivially God would not exist, hence God is dependent (in this sense) on himself.

The only workable strategy in the vicinity seems to be that of claiming that what exists necessarily cannot be created. But even this strategy will be blocked for those (and there are many of them) who think that a number of things other than God exist necessarily (in the sense that they are created, all right, but still God *had to* create them).

Department of Philosophy,
University of California,
Irvine, U.S.A.

GORDON G. BRITTAN, JR.

THE CARTESIAN COGITOS

The problems in ontology and the philosophy of logic on which Karel Lambert works are not new. They date from the very beginnings of Western philosophy. But his work provides original solutions to several of them, and in the process illuminates the history of their development. Much of my own work, in fact, has simply involved applying Lambert's results to 17th and 18th century texts.

In this paper I am going to apply two characteristically Lambertian views — his analysis of the presupposition relation and his associated claim that logic *per se* should be free of existential presuppositions — to the various texts in which Descartes argues for his own existence and which collectively are referred to as 'the Cogito'. My main claim will be that the Cogito is an *argument* and that it is plausible, if not also demonstrably sound. In several papers Lambert has commented briefly on Descartes, not always favorably. I hope he will not be too displeased by the way in which they have been made to join hands here.

Briefly, the argumentative or inferential interpretation of the Cogito is that 'I am' is somehow inferred (or inferable) from 'I think' in such a way that whatever certainty the latter enjoys is transferred to the former. On this reading, the Cogito has the structure of an argument or proof, a conclusion following from well-established premises. Slightly more precisely, the inferential interpretation has two parts: on the one hand, it tries to make definite the nature of the inferential *connection* between the two propositions or statements 'I think' and 'I am', and to provide evidence that they are so connected in Descartes' mind, and, on the other hand, it tries to make clear the nature of the *certainty* of the 'I think' premise which, by way of the inference, is transferred to the 'I am' conclusion.

Now to begin with, there are familiar passages in the *Discourse on Method* (AT VI, 33; CSM I, 127)[1] and the *Principles of Philosophy* (AT VIIIA, 7; CSM I, 194—95) which directly suggest the inferential interpretation, linking 'I think' to 'I am' with an emphatic 'therefore', Moreover, the inferential interpretation is invariably suggested in the various letters in which Descartes speaks of the Cogito, for example to

Wolfgang Spohn et al. (eds.), Existence and Explanation, 25—38.

Reneri for Pollot, April, 1638 (AT VI, 35)[2] and again to Colvius, November 14, 1640 (AT III, 247; Kenny, 84). Perhaps the most striking piece of textual evidence is to be found in the fifth set of Objections to the *Meditations* where Gassendi reconstructs the Cogito as an inference (AT VII, 258—59; CSM II, 180) and to which, in his Reply, Descartes agrees (AT VII, 352; CSM II, 243—44)

But anyone who advances the inferential interpretation must not only gesture in the direction of such passages, however important they are in making a *prima facie* case, but also make clear the precise inferential structure of the Cogito and the 'indubitable' status of the 'I think' premise.

The line I will take is suggested most directly by the passage from the *Discourse* already referred to: "Immediately I noticed that while I wished thus to think that all was false, it was necessarily required that I who thought it was something." The most natural way to read this is as follows: the (certain) truth that I wished to think that all was false necessarily required, as a condition of its possibility, that I who thought it was something; in a word, that the truth of 'I think' *presupposes* an existent I. But to say this is to say, at least in part, that if 'I think' is true, then 'I am' must be true; in the general case, if any (singular, atomic) statement or proposition of the form 'Fa' is true, then '$(Ex)(x = a)$' is true, i.e., a exists. This is what Descartes means when he says in the *Discourse* that my existence is 'necessarily required' by my attempt to think that all was false. From the truth of 'I think' and this semantic principle, 'I am' follows.

In its most general form the principle is one way of taking the traditional dictum: what does not exist has no properties (*non entis nulla sunt attributa*). I will call this dictum the '*non entis*' principle. It is a principle to which Descartes makes frequent reference. Thus the *Principles of Philosophy*, Part I, Principle XI: "nothingness possesses no attributes or qualities. It follows that, wherever we find some attributes or qualities, there is necessarily some thing or substance to be found for them to belong to"; (AT VIIIA, 8; CSM I, 196). He takes it as a 'common notion' or logical axiom, on a par with the principle of non-contradiction. In this passage he is also quite explicit concerning its role in the Cogito reasoning.

It should be noted that the semantic principle, if any (singular, atomic) statement or proposition of the form 'Fa' is true, then '$(Ex)(x = a)$' is true is not simply another way of expressing the '*non entis*' principle. For the '*non entis*' principle, taken literally, says that if a given object *a* does not exist, then there is no property that *a* has. That is to say, it is a principle about objects and properties expressible only in a language that quantifies over properties. On general philosophical grounds, one might wish to avoid quantifying over properties. More to the point here is that Descartes' text very well fits my reconstruction.

Further light on the character and role of the '*non entis*' principle as I have construed it is cast by the arguments of commentators who have taken note of, then rejected or qualified, the possibility that the Cogito turns on some sort of connection between truth and existence. The most important of these arguments, Jaakko Hintikka's,[3] implicitly takes the form of a dilemma: if the inference from true predication to existence is valid, it is only because we assume from the outset that all singular terms refer, in which case Descartes' argument would be trivial or question-begging (i.e., since we would assume from the outset that the putative singular term 'I' referred); if, on the other hand, we admitted non-referring singular terms into the language in which we formulated the statements or propositions in question, then the inference would be invalid (although no longer trivial), witness the fact that it does not follow from 'Hamlet thought a great many things' that Hamlet existed. In brief, on this version of the inferential interpretation the Cogito is either question-begging or invalid. We would do well, therefore, to look for a more plausible alternative.

Hintikka's argument raises rather difficult issues about the nature of predication that are both crucial to our understanding of the Cogito and very difficult to resolve. But without attempting to resolve these issues for the moment, two rather serious questions can be raised about the argument itself.

Hintikka claims both that particularization, and the corresponding rule of existential generalization, fail (to wit, when non-referring terms are admitted into the language in question) and that the validity of the inference from 'I think' to 'I am' presupposes that all singular terms in the language refer. I believe that the first claim is misleading and the second false. At the very least, the second horn of his dilemma is to be rejected.

Suppose that we allow non-referring singular terms such as 'Hamlet' or 'Heimdal' into our language. What truthvalue do we accord the (singular, atomic) propositions or statements in which they occur? For those who are suspicious about Hamlet's thinking anything, it is enough to assert his self-identity, viz. the truth of 'Hamlet = Hamlet'. In Hintikka's view, such assertions are logically true, hence true, yet it doesn't follow (in the case where the singular terms flanking the identity sign fail to refer) that the objects need exist; '$a = a \rightarrow (Ex)(x = a)$' does not in general hold. For Hintikka, every statement of the form '$a = a$' is true, in particular, 'Hamlet = Hamlet', quite independent of whether *a* refers or Hamlet exists. But at least two other positions are possible: either statements of the form '$a = a$', in the case where *a* does not refer, are truthvalueless or they are false. If we claim that they are false, then we *can* hold that whenever '$a = a$' is true, *a* refers. Such a view characterizes what has come to be called the position of the 'negative free logician'.[4] Thus it is possible, *contra* Hintikka, to admit non-referring singular terms into the language, and to grant the failure of particularization in some complex contexts, e.g., 'Hamlet does not exist', and yet at the same time hold that the inference from 'I think' to 'I am' is valid, i.e., that the corresponding conditional is logically true, because 'I think' is an atomic, singular statement.

So much for Hintikka's conclusion. My second question concerns his premises. It is not enough to cite 'Hamlet thought many things' as a truth (from which the existence of Hamlet does not follow), for this would merely beg the question against the '*non entis*' principle as I have reconstructed it. Indeed, Descartes, who tended to equate falsity with 'mere fable' and was firmly committed to the correspondence theory of truth, would reject it outright. But what then about such apparently less controversial cases as 'Homer was either a Greek or a barbarian (non-Greek)'? If we make a distinction between predicative and non-predicative contexts, and hold that the '*non entis*' principle is restricted to predicative contexts, and take '. . . or not . . .' as a non-predicative context, then there is no difficulty. And this, indeed, seems to have been just Descartes' position. The extremely important Principle XLIX of Part I of the *Principles of Philosophy* reads:

> Everything in the preceding list we regard either as a thing or as a quality or mode of a thing. But when we recognize that it is imposible for anything to come from nothing, the proposition *Nothing comes from nothing* is regarded not as a really existing thing, or even as a mode of a thing, but as an eternal truth which resides within our mind. Such truths are termed common notions or axioms. The following are examples of this class: *It is impossible for the same thing to be and not to be at the same time; What is done cannot be undone; He who thinks cannot but exist while he thinks*; and countless others the whole list of which it would not be easy to ennumerate (AT VIIIA, 23–4; CSM I, 209).

The law of excluded middle, here in the form of the law of non-contradiction, is on this list of 'common notions'. Just as clearly, according to Descartes, it is not the mode of a thing, in our words cannot be predicated of a thing. It is, we might say, a merely conceptual truth, holding generally and therefore of nothing in particular. But as a result 'Homer was either a Greek or a barbarian' does not have the (properly predicational form) '*Fa*', hence the '*non entis*' principle cannot be applied to it. Descartes did not, I am reasonably sure, contemplate the possibility envisioned by the 'negative free logician' that 'Homer was either a Greek or a barbarian' may be false if 'Homer' does not refer. 'Homer was either a Greek or a barbarian' has whatever 'truth' the eternal verities, common notions, and mathematical axioms enjoy, whether 'Homer' refers or no. But this 'truth' is quite independent of any existential assumptions.

None of this is to say that Hintikka is simply mistaken about the larger philosophical issues concerning truth and predication. It is to say that Hintikka's argument fails to undermine the possibility of our reconstruction of an inferential intepretation of the Cogito, and that there is ample reason in the text for attributing just such a reconstruction to Descartes.

The 'I think' premise must now be examined somewhat more carefully, although at the risk of seeming naive I think the insight involved is much less involved than it is often taken to be. Perhaps some principle of hermeneutic charity has been operating.

There are two (roughly equivalent) ways in which to contrast 'I think' with other propositions or statements Descartes dismisses as certain truths in the first Meditation. On the one hand, whereas the dismissed propositions all rest in one way or another on inferences, the proposition 'I think' does not. First principles ('absolute simples'), among which 'I think', are those which, although they serve as premises in derivations, are not themselves derived. Similarly in the second Meditation, where mathematical propositions are also called in doubt, it is the unreliability of any inference to a world (given the possibility of a *malin genie*) to which these propositions apply that makes us suspect their *truth*. But again, the truth of 'I think' does not depend on an inference, reliable or not. Descartes often puts this *immediate* character of 'I think' by saying that it is 'intuitive'.

On the other hand, none of the dismissed propositions in the first or second Meditations is either self-intimating or incorrigible[5] in the way that 'I think' is.[6] A proposition *p* is incorrigible for a particular person *A* just in case

(1) *A* believes that *p*

(2) (*A*'s belief that *p*) logically implies that *p*

Descartes seems to want to say that propositions about current states of mind are incorrigible in just this sense: if I believe that I am thinking (perceiving, believing, etc.), then it follows logically that I am thinking, etc. I cannot be mistaken about my own mental states. *P* is self-intimating for *A* just in case

(1) *p*

(2) (*p*) logically implies (*A* believes that *p*)

Again, if *p* is a proposition about a current mental state, then I believe that *p*. There are no unnoticed mental states (i.e., Descartes does not allow for the possibility of *petites perceptions*).

Applying these two notions to current mental states, it follows that I cannot be in error or ignorance about them. If I have a particular thought, then I know (am aware + cannot be mistaken) that I have it. States of mind are in this way completely 'transparent', in contrast to objects, which are opaque.

Now in virtue of these considerations, 'I think' is certain, i.e., knowably true. It cannot be doubted, because doubt is possible only when an inference has been made; but in the case of claims about our own mental states there is no inference (no 'evidence', etc.). It cannot be doubted, in a second place, because doubt is possible only where there is the possibility of error; but I cannot be mistaken about my own mental states — they are as they appear and they appear as they are.

Reconstructed, the argument of the Cogito is thus as follows:

(1) I am certain that I think, i.e., 'I think' (alone among propositions so far canvassed) must be true.
(2) But a true (singular, atomic) proposition or statement presupposes the existence of the object to which its subject term refers.
(3) 'I' is the subject term of the true proposition in question.[6]
(4) Therefore, I exist, i.e., this proposition must be true.

It is often thought that Descartes' formulation of the Cogito in the *Meditations*, unlike that in either the *Discourse* or the *Principles* in its omission of the *cogito, ergo sum* formula, particularly resists an inferential interpretation. Before proceeding further, we might see how our reconstruction holds up as a reading of that text.

The formulation in the second Meditation is somewhat more complicated than elsewhere.[7] It can be analysed into four distinct, although inter-related stages. I will quote, then note, the salient features of each in turn, and finally reconstruct the main lines of the arguments.

(1) I have just convinced myself that nothing whatsoever existed in the world, that there was no sky, no earth, no minds, and no bodies; have I not thereby convinced myself that I did not exist? Not at all; without doubt I existed if I was convinced (French version: ". . . or even if I thought anything").

The following points can be noted. This argument is in the past tense, it is hypothetical ("If I was convinced, . . . , then without doubt I existed") in form, and its strategy is that of a *reductio ad absurdum*. On this formulation of the argument, the presupposition relation is illustrated, but the premise is not indubitable. Could it, in fact, ever be merely *true* that I convinced my self that I did not exist?

(2) Even though there may be a deceiver of some sort, very powerful and very tricky, who bends all his efforts to keep us perpetually deceived, there can be no slightest doubt that I exist, since he deceives me: . . .

This argument is in the present tense, it is hypothetical (if the demon deceives me, then there can be no slightest doubt that I exist) in form, and again its strategy is that of a *reductio*: the supposition of an all-deceiving deceiver leads to the conclusion that I cannot be deceived about my own existence. It is also what we might call an 'object' formulation of the Cogito; the self enters the argument as the object of perception, not as the subject of thought (or doubt). Finally, on this formulation the presupposition relation is illustrated (in this sense, that if it is true that the demon deceives me, then both the demon and I must exist), but the premise is not 'indubitable'. Indeed, it turns out that the premise is false since, on Descartes' argument in the third Meditation, the existence of such a demon is incompatible with the existence of a benevolent God.

(3) and let him deceive me as much as he will, he can never make me be nothing as long as I believe that I am something.

The argument is hypothetical (if I think that I am something, then it is not possible that I am nothing), although the strategy this time is direct rather than indirect. On this formulation, the presupposition relation is illustrated, but no reason is given for thinking that the premise is *true*, much less that it is indubitable.

(4) Thus, after having thought well on the matter, and after examining all things with care, I must finally conclude and maintain that this proposition: *I am, I exist*, is necessarily true every time that I pronounce it or conceive it in my mind.

This argument is hypothetical (if I pronounce or conceive the proposition 'I am, I exist' then on that occasion the propsition is necessarily true) and the strategy is direct, not indirect. For the fourth time, the presupposition relation is illustrated, and this time a reason is given for thinking that the premise must be true; i.e., if I am thinking that I exist, I must be thinking.

If this is a more or less accurate summary of the argument-stages in the second Meditation version of the Cogito, then the following is true. First, every formulation is hypothetical in character, and its validity depends on the '*non entis*' principle. Second, there is a regular development in these formulations in that we go from a supposition that is in some sense 'impossible', to one that is possible, but in fact false, to one that is true, but uncertain, to one that is true and certain.

From a slightly different, and still general, point of view, it seems to me that the over-all argument of the second Meditation has the following structure: Either I am deceived in thinking that I exist or I am not deceived. If I am deceived in thinking that I exist, then either I deceived myself or someone else (a very powerful demon) deceives me. I couldn't have deceived myself that I exist, because the truth of 'I deceived myself' presupposes the existence of an 'I'. The demon can't deceive me that I exist, because the truth of 'the demon deceives me' presupposes the existence of a 'me'. But if I am not deceived in thinking that I exist, then I must be correct in thinking that I exist. Therefore, quite certainly, I exist. "Thus (concluding on the basis of the foregoing), after having thought well on the matter, and after examining all things with care (i.e., after systematically examining all possible cases), I must finally conclude . . ."

This reconstruction of the argument is, I think, alright as far as it goes, but we need a tighter focus. Notice two points in particular. The second and third steps in the argument are linked: if the demon deceives me about my own existence, then I must have the false thought that I exist. But that I have a false thought entails that I have a thought.

And if it is true that I have a thought, then, by the '*non entis*' principle, it follows that I exist. That is, it is misleading to say that the argument at the second stage goes 'the demon couldn't deceive me without there being a *me* to deceive'. Rather, we are to assume immediately the point of view of the subject (his thoughts), from which point of view it is not possible that I am nothing given that I think anything at all. Otherwise, we *might* be tempted to argue (mistakenly, since the antecedent is uncertain), 'if the demon curses me, then I exist', because 'the demon curses me' couldn't be true unless I existed.

The third and fourth steps are also linked. The fourth step makes explicit that the structure of the Cogito involves a relation between propositions or statements and provides an argument for the supposition of the third step.[8] The conclusion is that the proposition 'I exist' is true. The argument is that if I think I exist, then I must think something. That is, nothing turns on the fact that the proposition thought has to do with my existence.

But if I think anything, then I must exist. In other words, the truth of the proposition 'I think' presupposes the truth of the proposition 'I exist'. Since I cannot be deceived that I think, I cannot be deceived that I exist. The proposition 'I exist' is not simply *true*, but *certain*.

Questions remain. We need a still tighter focus on the fourth step or stage in the argument, in particular on the assertion that "'*I am, I exist*', is necessarily true every time that I pronounce it or conceive it in my mind." Hintikka takes this passage as the chief support of his so-called 'performative' interpretation of the Cogito. On this interpretation, the *act* of pronouncing or conceiving 'I exist' *verifies* it (i.e., the assertion that I exist is self-verifying). As a result, Hintikka emphasizes the 'momentary' character of the Cogito in the second Meditation.

On my reading, there is what seems to me to be an equally plausible interpretation that avoids the many difficulties of the 'performative' reconstruction, among others that 'cogito' has to be limited to a narrow range of mental states and that the proposition cogitated has to be 'I exist', neither compatible with a number of passages in which the Cogito is set out. One aspect of the alternate reading has to do with the inferential connection, the other with the 'indubitable' character of the premise.

To this point, and in a rather deliberate way, I have talked about 'statements or propositions' as comprising the steps of the argument. But in fact the bearers of truth are *statements*. But statements are dated utterances (or their purely mental equivalents). Thus the statement 'I think' (pronounced or conceived) presupposes as a condition of its truth, *at the time it is made*, that I exist.

Moreover, the certainty of the 'I think' premise depends, or so I have suggested, on its self-intimating and incorrigible character. Thus the premise is certain on all and only those occasions on which I am in fact thinking. As befits the summary character of the fourth step of the argument in the second Meditation, Descarates runs this point together with that made in the last paragraph.

Perhaps it would help both to clarify and support my reconstruction of the Cogito to draw attention to some neglected historical material. On my reconstruction of the argument of the second Meditation in particular, there are *two* Cogitos, one of which takes the form of a *reductio*, the other of which (as in the *Discourse* and the *Principles*) is direct.

The *reductio* form of the Cogito is the so-called *consequentia mirabilis*, in a modern version $(\sim p \supset p) \supset p$.[9] Four points might be made in connection with this historically important, but inadequately studied argument form. First, it is one of the few propositional, as opposed to syllogistic, forms studied by Aristotle.[10] Second, in Aristotle, and especially among the Stoic logicians for whom the pattern was central, it (together with *reductio* properly so-called) is used to refute the sceptics by showing that the sceptical position (e.g., that all truths are relative, that nothing can be proved, and so on) must be false since it entails its own negation (or, the other way around, that the anti-sceptical position must be true since entailed by its own negation). Third, this is the pattern used by St. Augustine in a passage which very much resembles the Cogito and which may or may not have directly influenced Descartes.

> In respect of these truths, I am not at all afraid of the Academicians, who say, What if you are deceived? For if I am deceived, I am. For he who is not, cannot be deceived; and if I am deceived, by this same token, I am. And since I am if I am deceived, how am I deceived in believing that I am? for it is certain that I am if I am deceived. Since, therefore, I, the person deceived, should be, even if I were deceived, certainly I am not deceived in the knowledge that I am.[11]

It follows that Augustine precurses Descartes in the form, purpose, and strategy of the reductio version of the Cogito, as a way of silencing the sceptics. There is, on the other hand, no direct version of the Cogito in

Augustine's writings (so far as I know), and although he too insists on the incorrigibility of claims about one's own mental states, there is no attempt to link this fact with the certainty of one's own existence. Descartes, like Augustine, thought that the sceptical position was self-defeating, but he also thought that in the Cogito one's existence was self-revealed. Fourth, the *consequentia mirabilis* occurs (although not so labled) in Euclid[12] and, more importantly for our purposes, in the 'modern Euclid', P. Clavius (1537—1612), whose textbook of geometry was standard fare in Jesuit schools and which furnished, with little doubt, Descartes' introduction to the subject.[13] A careful reading of the second Meditation, then, shows how Descartes, in what I have called its first two argument-stages, made use of the form and perhaps also the tradition in turning the tables on the sceptics.

Why did Descartes set the direct proof of the second two argument-stages of the second Meditation alongside the indirect proof just discussed? I suggest that it was because the indirect proof was strategically, but not also philosophically adequate. In this particular case, the premises of the indirect proof were 'impossible' and false respectively, whereas Descartes held that the 'I think' premise of the direct version of the argument was certain, and in a very significant way. More generally, Descartes held that *reductio* proofs were to be replaced whenever possible by direct proofs. I know of no text where he is explicit about this (although *reductio* proofs are notably absent from his *Géométrie*), but in Part IV (which he credits to Descartes), chapter 9 of *The Art of thinking*, 'Some defects commonly encountered in the geometer's method', Arnauld says of such demonstrations that they

> may convince the mind, but they do not enlighten it; and enlightenment ought to be the principal fruit of our knowledge. Our minds are unsatisfied unless they know not only *that* a thing is but *why* it is. But the *why* is never learned from a demonstration which employs a *reductio ad absurdum*.[14]

Finally, I want to raise two larger questions: why did Descartes think that he could *prove* that he existed and why was this proof of interest to him? Why, indeed, should anyone be interested in a proof of his or her own existence when, in a very trivial way, that existence is so manifest?

The first question involves others, so far delayed, concerning the status of the '*non entis*' principle. Granted, as I think is necessary, that there is something both self-intimating and incorrigible about ' I think', i.e., that there is a fundamental sense in which the premise of the Cogito argument is 'indubtitable', what is it, if anything, about the '*non entis*' principle that resists the corrosion of doubt? It isn't enough to say, as Descartes does occasionally, that the principle is an eternal truth or that it is perceived by the natural light. Nor is it enough to say that without this principle we couldn't prove the existence of anything, for at this point in the discussion it is presumably an open question whether anything exists.

There are, in fact, two questions involved at this point. One is whether doubt can legitimately be raised about this *sort* of principle. The other is whether this particular principle is true.

The answer to the first question seems to me to be 'no'. A view about the nature of predication, which is what the '*non entis*' principle amounts to, should be separated epistemologically from views concerning the way the world is, even if the former views turn out to have ontological implications. Whatever one might wish to say about Descartes' use of 'common notions' or axioms, the fact remains that the kind of evidence which might be taken to support their introduction differs from the kind of evidence used to substantiate claims about the world. Moreover, the conduct of any method resembling Descartes' inevitably rests in an obvious way on certain logical and linguistic presuppositions which must be satisfied if successful communication is to take place. Descartes and his readers have to agree, at least implicitly, on the meanings of the words used, the syntactic and semantic rules invoked, and so on, before sceptical doubts along the lines of the first Meditation can be raised or the arguments leading to them sketched and understood. These logical and linguistic presuppositions cannot, therefore, be very well *themselves* subject to such sceptical doubt. We have first to understand what the sceptic claims and why.

The answer to the second question, concerning the correctness of the '*non entis*' principle, is much more difficult. It does not follow from the fact that hyperbolic doubt cannot coherently be raised with regard to it, that it is uncontroversial, still less that it is true. One of Lambert's great merits is to have raised serious questions regarding standard developments in logic that were long taken as 'axiomatic'. The non-standard developments he helped pioneer not only made these questions serious, they also made clear that certain classical semantic principles (bivalence and the restriction to non-empty domains, for example) had ontological commitments which were to that point largely unrecognized. But these non-standard developments also make questions concerning the adequacy of particular theories of predication more difficult, if only because we now have more theories from which to choose. What, then, of the truth of the '*non entis*' principle, that the truth of a simple statement implies that its constituent singular terms always refer to existents? I am not at all sure. Semantic developments which reject this principle, many of them inspired by Meinong, tend either to make 'exist' equivocal or to allow large-scale expansion of their ontologies. Neither course is desirable, but I don't have an argument to prove the point.

The last question, why was Descartes interested in a proof of his own existence, is equally troublesome. Of course, Descartes wants to show that at least one proposition is, in context, 'indubitable' and provides us with a metaphysical handhold. But it has always struck me as paradoxical that Descartes uses the Cogito to establish an existence claim, the first in the order of discovery, but that this claim (it turns out), however certain and secure, has nothing to do with what exists *in the world*. It does not involve an inference from thought to what exists independent of thought. That inference is made only when we come to prove that God exists. The Cogito does not dispel any of the doubts raised by the first Meditation, nor, for that matter, any of the doubts raised by the possibility of a *malin genie*. At most we can say that the concept of an all-deceiving deceiver is, in one important respect, incoherent. Undoubtedly this is part of the point: the question of my existence can be settled apart from and prior to settling any questions concerning the existence and nature of the world. Perhaps this also provided Descartes with a reason for saying that in some sense 'I' am not *in* the world.

At a deeper level, the Cogito might be necessitated more by an ontological than an epistemological consideration, the fact that neither intentionality nor incorrigibility, the two main marks of the mental, serves to characterize the set of 'states of consciousness' which we might otherwise identify with the self. They serve only to mark out the sapient and the sentient respectively. It is, to the contrary, simply because they are *my* (or *your*) states of consciousness, attributes of an 'object' (the 'subject' of consciousness) whose existence has already been demonstrated, that they constitute a *set* and in terms of which they have any kind of inner unity.[15]

Montana State University,
Bozeman, U.S.A.

NOTES

[1] I refer in the usual way to the *Oeuvres de Descartes*, Adam and Tannery (eds.), revised edition, Vrin/C.N.R.S., 1964—76 and to the translation which is now standard in English, *The Philosophical Writings of Descartes*, Cambridge University Press, 1985 Cambridge, by Cottingham, Stoothoff and Murdoch, although the translations are not invariably from the latter.

[2] In *Descartes: Philosophical Letters*, translated and edited by Anthony Kenny, Clarendon Press, London, 1970, p. 52.

[3] In his classic article, '*Cogito, Ergo Sum*: Inference or Performance?' *Philosophical Review* **LXXI** 1962, pp. 3—32, and widely reprinted.

[4] See Karel Lambert, 'On the Philosophical Foundations of Free Logic', *Inquiry* **24** 1981, pp. 147—203. For examples, see Ronald Scales, *Reference and Attribution*, Ph. D. dissertation, University of California, California, 1969, and Tyler Burge, 'Truth and Singular Terms', *NOUS* **8** 1975, pp. 309—25.

[5] I borrow the terms and their explication from David Armstrong, *A Materialist Theory of the Mind*, Routledge and Kegan Paul, London, 1968, pp. 100ff., although the points are familiar in Descartes commentary. It is important to note that neither self-intimation nor incorrigibility turns on considerations of self-reference, the red herring in traditional discussions of the Cogito.

[6] Or any proposition about one of my mental states. Descartes uses the term *cogitatio* so widely as to cover virtually all of them.

[7] See H.G. Frankfurt, 'Descartes' Discussion of His Existence in the Second Meditation', *Philosophical Review* **LXXV** 1966, pp. 329—56.

[8] The fourth step supports the third, just as the third step supports the second. Gueroult, whom one criticises reluctantly, claims that ". . . le texte de la *Seconde Meditation* ne présente pas trace de raisonnement." *Descartes Selon l'Ordre des Raisons*, Aubier-Montaigne, 1968, I, p. 307. In fact, it seems to me one of the most closely argued passages in the history of philosophy.

[9] Whitehead and Russell, *Principia Mathematica*, #218. *Reductio ad absurdum*, properly so called, is the closely allied principle $(p \supset \sim p) \supset \sim p$, *PM*, #201. In what follows I draw upon William and Martha Kneale, *The Development of Logic*, Oxford University Press, Oxford 1962, pp. 96—7, 172—75, 346—49, 380—81, 524—25; Jan Łukasiewicz, *Aristotle's Syllogistic*, second edition, Clarendon Press, Oxford, 1957, pp. 50, 80; A.N. Prior, *Formal Logic*, second edition, Clarendon Press, Oxford, 1962, pp. 29, 51, 244.

[10] *Prior Analytics*, ii. 2 (53^b12) and ii. 4 (57^b6).

[11] *The City of God*, translated by Marcus Dods, The Modern Library, 1950, Book XI, #26, p. 370.

[12] *Elements of Geometry*, ix, 12.

[13] See Etienne Gilson, *Discours de la Methode; Texte et Commentaire*, 5th edition, Vrin, 1976, pp. 127—28.

[14] *The Art of thinking*, translated by J. Dickhoff and P. James, Bobbs-Merrill, 1964.

[15] I am much indebted to Margaret Wilson's account of the Cogito in *Descartes*, Routledge and Kegan Paul, London, 1978. My thanks to James Allard, Karel Lambert, and Jules Vuillemin for helpful comments on earlier drafts, and to Karel Lambert for twenty years of inspiration, encouragement, and friendship.

CHARLES B. DANIELS

UNDEFINED DEFINITE DESCRIPTIONS

Bertrand Russell [8] and John Bacon [1] argue that definite descriptions are not what they seem — referring terms on a par with names. Instead, they propose contextual definitions which serve to eliminate them. Russell analyses the sentence 'the *F Gs*' as 'there is at most one *x* such that *Fx* and there is at least one *x* such that *Fx* and *Gx*'. Bacon analyses it as 'for all *x* if *x* uniquely *Fs*, then *Gx*'. In either case all sentences containing definite descriptions are true or false even when there fails to be a unique entity described. For Russell 'The present King of France is bald' is false, for Bacon true. Bacon's analysis, in particular, meshes nicely with a feature of generalizations: 'All present Kings of France are bald' is also true when there are no present Kings of France.

Strawson [9, 10] argues (Frege [5] as well) that definite descriptions are referring terms on a par with names, and that when they lack a referent, sentences containing them are neither true nor false. Free logics capture the intuitions of those who, like Strawson, hold that reference failure leads to lack of truth value.[1] Of course, even in free logics definite descriptions *can* be contextually defined in the manner of Russell or Bacon. The motivating idea, however, is to treat the definite article as a primitive operator that forms singular terms from well-formed formulae. 'The present King of France' ('the *x* such that *x* is a present King of France') is treated as a singular term, which, like the name 'Pegasus', has no actual referent. When a singular term lacks a referent, most sentences containing it fail to have truth value.

However, 'Pegasus is brown or it's not the case that Pegasus is brown', 'Pegasus is Pegasus' and '($[\forall x]Fx \wedge$ Pegasus exists) $\supset F$ Pegasus' are theorems despite the presence in them of the non-referring term 'Pegasus'.[2] In free logics sentences are evaluated and then subject to *super*valuation.[3] Sentences like these are true on all supervaluations.

The Strawsonian view is intuitive and can be attractive, but what is often not fully appreciated is the intimate connection there is among primitive definite descriptions, the substitutivity of identicals, and necessity. An argument Donald Davidson used long ago in another context[4] can be adapted to expose this.

Wolfgang Spohn et al. (eds.), Existence and Explanation, 39—44.

Consider a presuppositional logic with primitive definite descriptions, identity, and necessity. Where 'E' is an existence predicate, the variable y does not occur in the wff A, no variables occur free in the wff B, and $Ac//t$ is the result of replacing zero or more occurrences of the singular term c in the wff A by the singular term t, the following seem plausible theorems:

(1) $(A \wedge c = t \wedge Ec \wedge Et) \supset Ac//t.$

(2) $E((\iota x)A) \equiv (\exists y)(\forall x)(A \equiv x = y)$

(3) $[B \wedge (\exists y)(\forall x)(A \equiv x = y)] \supset (\exists y)(\forall x)[(B \wedge A) \equiv x = y].$

(4) $[E((\iota x)A) \wedge (\forall x_1) \ldots (\forall x_n)(A \equiv C)] \supset [(\iota x)A = (\iota x)C],$
where $x_i, \ldots, x_n$ are all the variables free in A and C.[5]

From the above it follows that

(5) $[B \wedge E((\iota x)A] \supset [(\iota x)A = (\iota x)(B \wedge A)].$

We have then

(6) $[B \wedge E((\iota x)A)] \supset E((\iota x)(B \wedge A))$

(7) $[\Box(c = c) \wedge B \wedge c = (\iota x)A \wedge Ec \wedge E((\iota x)A)] \supset$
$[c = (\iota x)(B \wedge A)]$

(8) $[\Box(c = c) \wedge B \wedge c = (\iota x)A \wedge Ec \wedge E((\iota x)A)] \supset$
$\Box(c = (\iota x)(B \wedge A))$

(9) $[\Box(Ec \supset Ec) \wedge B \wedge c = (\iota x)A \wedge Ec \wedge E((\iota x)A)] \supset$
$\Box(Ec \supset E((\iota x)(B \wedge A))$

(10) $[\Box(Ec \supset Ec) \wedge B \wedge c = (\iota x)A \wedge Ec \wedge E((\iota x)A)] \supset$
$\Box[Ec \supset (B \wedge E((\iota x)A))]$

(11) $[\Box(Ec \supset Ec) \wedge B \wedge c = (\iota x)A \wedge Ec \wedge E((\iota x)A)] \supset$
$\Box(Ec \supset B).$

By (7) if Margaret Thatcher is Prime Minister and Venus is the Morning Star, then necessarily Venus is the x such that Margaret Thatcher is Prime Minister and x is a Morning Star. By (11) necessarily if Venus exists, then Margaret Thatcher is Prime Minister.

By distribution of $\Box$ over $\supset$, we also have from (11):

(12) $[\Box(Ec \supset Ec) \wedge B \wedge c = (\iota x)A \wedge Ec \wedge E((\iota x)A)] \supset$
$(\Box Ec \supset \Box B).$

Since $\Box A \supset A$ is a theorem, by strengthening the antecedent, we also have:

(13) $[\Box(Ec \supset Ec) \land B \land c = (\iota x)A \land \Box Ec \land E((\iota x)A)] \supset \Box B$

Five is the prime number between four and seven, and necessarily five exists. So necessarily Margaret Thatcher is Prime Minister.

These difficulties do not arise when definite descriptions are contextually defined. The reason is that when defined terms are eliminated, nothing of the form '$c = t$' contains a definite description. The substitution axiom has in its antecedent '$c = t$', where 'c' and 't' are *names* or *variables.* Neither can be a definite description. The moral here is that at least one of the following trio has to go: undefined definite descriptions, the substitutivity of identicals, or necessity.

In [3] the author has taken the Strawsonian view in a different direction. Instead of using supervaluations to make such sentences as 'Pegasus is Pegasus' be true, *all* sentences of classical first-order logic that contain non-referring singular terms lack truth-value, save sentences of the form 't exists' and 't doesn't exist'. Given the regular connectives $\sim$, $\supset$, $\land$, and $\lor$, this, of course, doesn't leave much to manufacture theorems with. What is used in most theorems as the primary connective is story implication, $\rightarrow$. '$A \rightarrow B$' is true when in all stories and the actual world, where A is true, B is true. Unlike the standard connectives, $\sim$, $\land$, $\lor$, and $\supset$ of classical logic, implications like '$A \rightarrow B$' can be true when antecedent, consequent, or even both lack truth-value.

As regards stories, there are at least three interesting features they possess:

(a) Not everything there is in every story. Neither Bertrand Russell or Captain Ahab are in the story *Oliver Twist.*

(b) Theorems may fail to be true in stories and may, indeed, be false. All sentences fo the form '$A \rightarrow A$' needn't be true in *Oliver Twist*; 'A' might be the sentence 'Bertrand Russell is not Captain Ahab'.

(c) Disjunctions may be true in a story in which neither disjunct is true. It may be true in a story that either Mary's mother is a spy or Mary's father is one, and yet the story may give no indication which parent is a spy.

(d) $(\exists x)Fx$ may be true in a story where for no term t is Ft true. It may be true that someone m urdered the heroine's father long before the story begins, without it being settled who did. In an admittedly frustrating mystery story, it might not be true that *everyone* has a perfect alibi and yet each character named or otherwise referred to in it does have one.

When primitive definite descriptions are introduced into a story logic, it turns out that a version of Davidson's argument can be used to show that if something from the actual world figures in a story, everything true in the actual world is true in the story. (Of course, *other* things may be true in the story as well.) For instance, if Nelson Algren's *Man with the Golden Arm* is set in Chicago, it is set in the birthplace of Charles B. Daniels, since it is actually true that I was born in Chicago. So by substitution of identicals I am a character in *Man with the Golden Arm*, and by considerations similar to those presented above so is Margaret Thatcher.

The alternatives here seem to be four: (i) Swallow the awkward consequence and say that while Thatcher and I are in *Man with the Golden Arm*, we are in the dull and ignorable part of it. (ii) Hold that while the action of *Man in the Golden Arm* takes place in a city named 'Chicago', it doesn't really take place in the Chicago that is my birthplace. Indeed, nothing in the fiction we are acquainted with is anything but fictional, although names are used ambiguously between the actual world and stories, just as they are within the actual world itself. Many people are named Charles Daniels, and there is more than one Berlin. (iii) Abandon the substitutivity of identicals. (iv) Define definite descriptions contextually and avoid the whole mess.

A closing word concerning (iii).

$$(1') \qquad (A \wedge \Box(c = t) \wedge Ec \wedge Et) \supset Ac/t$$

might be proposed instead of (1). Thus mere identicals are not, in general, freely substitutable. In free logics with necessity, there is no reason to believe that necessarily Venus is the Morning Star, much less that Venus is the x such that Margaret Thatcher is Prime Minister and x is a Morning Star. So the foregoing difficulties do not arise.

A view like this has a number of its own problems, I think:

(a) While identicals are not everywhere substitutable unless necessary, why isn't there a stronger relation than identity, namely, universal indiscernability (necessary identity), that things may have to themselves. Let '$=_d$' represent this relation. We clearly have $t =_d t$ and $(A \wedge c =_d t \wedge Ec \wedge Et) \supset Ac//t$. The number of discernible things there are, say, in a box may thus exceed the number of non-identical things in the box. What then is the nature of the things that are not only identical to themselves but indiscernable from themselves?

(b) Can we locate all actual things in all worlds? That is, the quantifier $(\forall x)$ in $(\forall x)A$ refers to all actual things. Thus, can we instantiate $(\forall x)\Box B$ to $\Box Bx/t$ for all things such that Et?

(c) Finally, if one is attracted to story implication and lets

$$(1'') \quad (A \wedge \Box(c = t) \wedge Ec \wedge Et) \rightarrow Ac//t$$

replace the substitution axiom, $\Box A$ can be interpreted as saying that A is true in all stories that are *consistent, maximal, and universal.* In all such stories '$Et \vee \sim Et$' holds for *all* terms t. But then if three exists in all consistent, maximal, and universal stories (and why not?), it will follow that necessarily three is identical with the x such that Margaret Thatcher exists or Margaret Thatcher doesn't exist and x is a prime number between two and five. So by substitution Margaret Thatcher will be in every story in which three exists, even ones that are not consistent, maximal, and universal — like, for instance, the story *The Three Bears.*

Department of Philosophy,
University of Victoria,
Victoria, Canada.

NOTES

[1] A thorough presentation of the history, motivation, and ins and outs of free logics is given by Karel Lambert in [6].

[2] Wittgenstein [15] found this *very* odd. If a nominative N can fail to refer, how is it that 'N exists' is always true or false? If 'N exists' is always true or false, then the 'N' in it must always refer. If in some ofter context 'N' fails to refer, the 'N' is not, he would argue, being used univocally.

[3] The idea of supervaluation is Bas van Fraassen's. For more on supervaluations and free logics the reader is referred to Van Fraassen [11, 12, 13], to Van Fraassen and Lambert [14], and to quantificational extensions by Meyer and Lambert [7] and Bencivenga [2].

[4] See [4], pp. 91—2.

[5] Instead of (4) one might use: (4.1) $E((\forall x)A) \supset (C \equiv (\exists y)((\forall x)(A \equiv x = y) \wedge C((\imath x)A)/y)$, where y occurs in neither A nor C and there are no free variables in $(\forall x)A$.

REFERENCES

[1] Bacon, John: 1965, 'An Alternative Contextual Definition for Descriptions', *Philosophical Studies* **16** 75–6.

[2] Bencivenga, Ermanno: 1981, 'Free Semantics', *Boston Studies in the Philosophy of Science* **47**, 31–48.

[3] Daniels, Charles B: forthcoming, 'Definite Descriptions', *Studia Logica.*

[4] Davidson, Donald: 1967, 'The Logical Form of Action Sentences', Nicholas Rescher (ed.), *The Logic of Decision and Action*, University of Pittsburgh Press, Pittsburgh.

[5] Frege, Gottlob: 1960, 'On Sense and Denotation', in Peter Geach and Max Black (eds.), *Translations from the Philosophical Writings of Gottlob Frege*, Basil Blackwell, Oxford.

[6] Lambert, Karel: 1981, 'On the Philosophical Foundations of Free Logic', *Inquiry* **24**, 147–203.

[7] Meyer, Robert K., and Lambert, Karel: 1968, 'Universally Free Logic and Standard Quantification Theory', *Journal of Symbolic Logic* **33**, 8–25.

[8] Russell, Bertrand: 1969, *Problems in Philosophy*, Oxford University Press, London, Oxford and New York.

[9] Strawson, P. F.: 1950, 'On Referring', *Mind* **59**, 320–44.

[10] Strawson, P. F.: 1952, *Introduction to Logical Theory*, Methuen, London.

[11] Van Fraassen, Bas C: 1966, 'Singular Terms, Truth-Value Gaps, and Free Logic', *Journal of Philosophy* **63** 481–95.

[12] Van Fraassen, Bas C: 1966, 'The Completeness of Free Logic', *Zeitschrift für Mathematische Logik und Grundlagen der Mathematik* **12**, 219–23.

[13] Van Fraassen, Bas C: 1969, 'Presuppositions, Supervaluations, and Free Logic', in Karel Lambert (ed.), *The Logical Way of Doing Things*, Yale University Press, New Haven and London.

[14] Van Fraassen, Bas C., and Lambert, Karel: 1967, 'On Free Description Theory', *Zeitschrift für Mathematische Logik und Grundlagen der Mathematik* **13**, 225–40.

[15] Wittgenstein, Ludwig: 1961, *Tractatus Logico-Philosophicus*, Routledge and Kegan Paul, London and New York.

DANIEL HUNTER

MAXIMUM ENTROPY UPDATING AND CONDITIONALIZATION*

1. INTRODUCTION

Maximum Entropy Updating, or MAXENT for short, is a method for revising probabilistic beliefs when new constraints on probabilities are given. It produces a new probability measure that in a certain sense 'minimally' revises the prior probability measure so as to accomodate the new constraints.

MAXENT is often thought of as a *generalization* of Bayesian conditioning. If the updating constraint consists simply of the assignment of probability one to a proposition, MAXENT produces the same posterior when applied to that constraint as does conditioning on the proposition. However, MAXENT seems to be applicable in cases in which Bayesian conditioning is not, namely in those cases in which some of the updating constraints assign propositions probabilities strictly between zero and one.

Brian Skyrms (1985, 1987), however, has recently stood this relation between MAXENT and conditioning on its head, arguing that MAXENT should be viewed as a special case of conditioning, valid only under certain specific conditions. On Skyrms' view the constraints given to MAXENT are taken to be propositions in some suitably enriched probability space and the desired posterior is produced by conditioning on these propositions. MAXENT is justified, on this view, only if it produces the same posterior on the original space as does conditioning on the constraints.

I agree with Skyrms that whenever MAXENT gives a valid result, conditioning within some larger space may be used to obtain the same result. I also agree that MAXENT may give the wrong result if naively applied to constraints of a certain type. But our agreement is superficial. For we disagree about what sort of proposition is an appropriate conditioning proposition; and, more importantly, we disagree radically about the relative priority of MAXENT and conditionalization. I view MAXENT as prior to, and justificatory of, a certain type of conditioning.

Wolfgang Spohn et al. (eds.), Existence and Explanation, 45—57.

To explain these diagreements is the task of this paper. First, though, I give a brief description of MAXENT as a purely mechanical tool for updating. Next, I examine Skyrms' objections to MAXENT and argue that they arise from using MAXENT in an epistemological setting for which it is not suited. Then I argue that given the right epistemological setting, the validity of MAXENT almost trivially follows. I conclude with an account of what I think is the correct relation between MAXENT and conditioning.

2. MAXENT

MAXENT takes a probability measure $P(\ )$ over a space $S = \{X_1, X_2, \ldots\}$ of mutually exclusive and exhaustive states (we confine our attention to countable spaces) and constraints on new probabilities for propositions over this space, and produces a posterior measure satisfying the constraints. As explained below, MAXENT makes the minimal revision necessary to the prior in order to accomodate the constraints.

A common form of constraint is one specifying the expected values of certain random variables. A constraint specifying a new probability for a proposition may be thought of as of this type by considering a random variable taking the value one if the proposition is true, and zero otherwise; the expectation of this random variable is just the probability of the proposition in question. For the purposes of this paper, we may suppose the constraints to be of this form.

MAXENT chooses from among the probability measures satisfying the constraints, a measure $Q(\ )$ such that the relative entropy $H(P, Q)$ is minimal, where $H(P, Q)$ is defined to be:

$$\sum_{i=1}^{\infty} Q(X_i)\log(Q(X_i)/P(X_i))$$

The minimum relative entropy posterior is a minimal revision of the prior measure in the sense that the information gain in going from the prior to the posterior is minimal among all measures satisfying the constraints.[1]

For conditions on the existence and uniqueness of the minimum relative entropy posterior, see Csiszar (1975).

3. SKYRMS' OBJECTIONS TO MAXENT

In (1985), Skyrms argues that MAXENT may be viewed as a way of conditionalizing on evidence, but a way that is valid only in special circumstances. His argument goes as follows. MAXENT performed on a space S with constraint C can be interpreted as conditionalizing on the constraint C in a larger space within which C is defined. For example, the constraint that the probability of heads on the flip of a particular coin is 0.6 could be expressed as a proposition in the space of possible sequences of flips of the coin, say as the proposition that the long run relative frequency of heads is 0.6. The constraint can then be assimilated simply by conditionalizing on this proposition within the sequence space.

The constraint might alternatively be interpreted as referring to a degree of belief, perhaps something like final belief, belief after the evidence is in. Skyrms (1980) shows that it is consistent to have a probability space in which final belief is a random variable.

As Skyrms points out, in order for MAXENT to agree with conditionalization in the enriched space, whether the enriched space refers to sequences of events in the smaller space or to degrees of belief about events in the smaller space, certain conditions must be satisfied. One such condition is *sufficiency*:

SUFFICIENCY: $P(H \mid A, C) = P(H \mid A)$

where C is a constraint on the new probability of proposition A.

However, says Skyrms, sufficiency is quite often violated. For example, if the constraint concerns a future degree of belief, then this degree of belief may have connections to certain propositions that what is believed does not. For instance, the probability that x will be worried given that his final degree of belief in his having cancer is 0.99 may be higher than the probability that x will be worried given simply that x has cancer.

Likewise, if probability is interpreted as physical probability or chance, there also seem to be cases in which sufficiency is violated. Skyrms (1985) gives this example: The probability that a coin comes from the U.S. mint given that it comes up head and that the physical probability of heads is 0.8 is lower than the probability that the coin comes from the U.S. mint simply given that it comes up heads.

In (1987), Skyrms further demotes MAXENT from its position as a sometimes valid rule of updating to something that is not a form of updating at all. Here he argues that MAXENT is a rule for *supposing*, not for updating. The argument he develops for this position is very interesting, but here I am more concerned with an example he uses in the course of developing the argument.

The example is this: Suppose we have a three-sided die that we believe is a trick die which always lands the same way, but we are unsure which is the favored side. Suppose we assign equal probabilities to the three ways in which the die might land. Then the expected number of spots equals two. Now suppose an experiment is run in which the die is tossed a number of times and it is reported that the average number of spots turned out to be two. When we give this information to MAXENT, what happens? The answer is 'nothing'. Since the constraint is *already* satisfied, MAXENT treats it as, in Skyrms' words, 'no news', and hence does not revise the probabilities. But, Skyrms argues, the information about the average number of spots is cause for revision. The probability of two spots should go from 1/3 to 1.

There is something very puzzling about this example. Is the constraint news or not? If it is not news, no valid updating rule should cause a change in belief. But if it is news, why is MAXENT forced to treat it otherwise? To solve this puzzle, we must place MAXENT in its proper epistemological setting, to which task we now turn.

4. MAXENT ON ITS OWN TURF

MAXENT is a method for revising belief states expressed in terms of probabilities. It takes two arguments, one a prior probability measure and the other a constraint on the posterior measure. The constraint should be viewed as just that, a *constraint*, not as a piece of information.

The distinction between a constraint and a piece of information is more clear-cut in the case of ordinary, non-probabilistic belief. If I come to believe a proposition *A* that I did not previously believe, the belief revision problem is that of modifying my beliefs so that they are compatible with A, but in such a way as to minimize unnecessary changes in belief. The proposition *A* serves to restrict the set of allowable belief states.

This is different from treating the new belief as a piece of evidence of the form 'I now believe *A*'. To do so is to be *self-reflective* in a manner not called for in ordinary, run-of-the-mill cases of belief revision. In fact, adopting such a self-reflective attitude may result in a final belief state in which A is not believed (e.g., I may distrust my first instincts).

Similarly, in the case of probabilistic belief, we should regard the constraint as expressing a certain change in one's belief state whose effect must be propagated throughout the system of beliefs. This is MAXENT's proper epistemological setting. We think of the initial belief change, by analogy with physical dynamic processes, as a perturbation that causes the belief system to be out of equilibrium and MAXENT's role is to restore (epistemic) equilibrium.

The distinction between a probability statement as a constraint and as a piece of information is particularly clear in Skyrms' die example. If we distinguish between the space S of possible outcomes of a single throw of the die and the space S^n of n trials of throwing the die, then the statement that the expectation of the number of spots is two can be seen to be ambiguous: it might be a statement about empirical averages in S^n; or it might be taken as a constraint on S. But these are two entirely different things, as Jaynes (1978) took great pains to make clear.

We could turn the tables on Skyrms by considering the following example: Suppose I have a uniform distribution over the possible chance models for the die. I throw the die and observe the outcome, but unfortunately the conditions of observation are not ideal. I am observing by dim candlelight and cannot quite determine how many spots are showing. Suppose my probabilities, based on the observation, for one, two, and three spots are 0, 0.5, and 0.5, respectively. If I took these probabilities as pieces of information about long run empirical averages, then I would assign a probability close to one to the chance model with those probabilities. But this is wrong. I cannot be that certain, just on the basis of that single observation, that that is the correct chance model.

Of course the supporters of conditionalization would cry foul here. What is learned is not information about long run averages, but only information about a single trial. But that's the point. A constraint on probabilities for a single trial is different from information about long run averages and MAXENT should not be faulted if it gives the wrong answer about S^n if it is forced to interpret the information as a constraint on S.

To be fair to Skyrms, it must be admitted that some supporters of

MAXENT have seemed to use it as a way of doing in S what conditionalization would accomplish in S^n, without being very careful about when the two techniques are guaranteed to agree. Jaynes (1963), for example, uses information about empirical averages for throws of a six-sided die as a constraint for MAXENT to produce a posterior measure over the space of outcomes of a single throw. But matters are more complicated than they at first appear. In (1978), Jaynes, responding to a flood of criticism over his die example, tried to make clear that he saw the problem purely as one of enforcing a constraint on S. True, the constraint was generated by information about S^n, but it was still a constraint on S, not on S^n. It is also relevant that Jaynes assumed we have no prior information about the space S^n, whereas in Skyrms' example we have a definite probability measure over S^n. I do not wish to defend all of Jaynes' applications of MAXENT, but it is important to recognize that Jaynes was aware of, indeed strongly emphasized, the difference between a constraint on S and a conditioning statement in S^n. Furthermore, the philosophical literature on MAXENT (e.g., Van Fraassen (1980), Domotor (1980), and Williams (1980)) has been much clearer about the proper epistemological role of MAXENT.

One must, however, concede to Skyrms that conditionalization over an enriched space is a powerful technique that can be used even if a precise prior over the higher-level space is not available. But the dispute is not over the merits of conditionalization when the conditioning information is properly construed as information about an event in the enriched space. The issue is whetehr or not uncertainty about the *single* case can properly be represented by embedding in a larger space. The next section attempts to show that there are dangers in such an approach.

5. HIGHER-LEVEL DANGERS

Suppose proposition A concerns some unique, non-repeatable event, so that a constraint on the new probability of A cannot be construed as a piece of information about repeated trials. There is still a way such a constaint might be represented as a proposition in a higher-level space. Let S be the space of ordinary, ground-level propositions (including A). We assume for simplicity that S is finite, of size n. Then we may consider the space T of probability measures over S. Given a probability measure PROB over T, we take the probability of a proposition in S to be its expected probability relative to PROB. Conditioning on a statement about the probability of a proposition in S then becomes straightforward: T can be identified with the set of points $(x_1, x_2, \ldots, x_n)$ of n-dimensional Euclidean space that satisfy $x_1 + x_2 + \ldots + x_n = 1$. A constraint on a probability for a proposition in S is equivalent to a constraint on a linear combination of the x_i. Conditionalizing on the probability for the proposition amounts to eliminating that portion of T that fails to satisfy the constraint on the linear combination of the x_i.

For example, let S be the space $(A \& B, A \& -B, -A \& B, -A \& -B)$, the joint space for the propositions A and B. Suppose that our probability measure over T is uniform. Then it is straighforward to show that our expectation of probability for S yields (0.25, 0.25, 0.25, 025) as the probabilities for the members of S. Note that A and B are probabilistically independent under this assignment of probabilities. Now suppose the probabilities for A and B change to 0.8 and 0.2, respectively, but we get no information about any of their joint probabilities. Taking the new probabilities as constraints on T and conditionalizing, we get this posterior for S: (0.1, 0.7, 0.1, 01.). But now A and B are no longer independent. $P_{\text{new}}(A) = 0.8$, but $P_{\text{new}}(A \mid B) = 0.5$. Yet there was nothing in the constraints to cause A and B to be dependent. Learning B should not lead to A's decreasing in probability.

If we accept that A and B should remain independent in the posterior and that the constraints are satisfied in the posterior, then the posterior must be (0.16, 0.64, 0.04, 0.16). There are measures over T that would yield this posterior over S, but it is far from obvious how such a measure might be arrived at independently of knowing what the posterior measure on S should be.

It might be suggested that the probability measure on T should concentrate all its weight on those points that make A and B independent. In that case, conditionalizing on the constraints would yield the right result. However, such a measure on T would prevent learning of a dependency between A and B, which we cannot rule out *a priori*.

MAXENT yields the right posterior over S when given the above constraints, and it is no accident that it does so. It is a general property of MAXENT that it preserves independence of propositions unless the constraints specify probabilistic relations between the propositions. This is something other proposed updating rules do not do (e.g., neither the minimization of Hellinger distance nor the minimization of variation distance, considered in (Diaconis and Zabell, 1982) as alternatives to MAXENT, satisfy this requirement).

Shore and Johnson (1980) give an axiomatic derivation of MAXENT. One axiom is in effect the requirement that the updating rule preserve independence under the conditions just described. Most of the other axioms seem equally unexceptionable, but one of them entails the controversial sufficiency condition. What the next section shows is that sufficiency is an almost trivial consequence of our description of the epistemological setting for MAXENT. The result proved in that section, when taken together with Shore and Johnson's axioms, should provide a more satisfactory foundation for maximum entropy updating than the latter alone.

6. SUFFICIENCY REVISITED

We wish to show that any reasonable updating function will satisfy the sufficiency condition. Let $P(\ )$ be a probability measure over space S and let $X_1, X_2, \ldots, X_k$ be a partition of S. Suppose the constraint C for updating involves only the partition elements — i.e. that the constraint is of the form $\Sigma_{i=1}^{k} s_i Q(X_i) = c$, where $Q(\ )$ is the posterior measure. We impose the following three conditions on any updating function $U(\ )$:

(1) $U(\ )$ is a function only of the prior measure over the space S and the constraint on the posterior (and so may be written $U(P, C)$).

(2) $U(P, C)(X_i) = U(P', C)(X_i)$ if P and P' agree on the partition elements of S.

(3) $U(P, C)$ is continuous in P.

The second condition is intended to capture the idea that the constraint reflects a change in belief regarding some field of propositions and that it is not necessary to consider refinements to this field in order to update. We expect agreement about the prior probabilities of the propositions involved in the change to yield agreement about their posterior probabilities.

Suppose now that constraint C assigns a new probability to proposition A. Sufficiency then says that $U(P, C)(H|A) = P(H|A)$. To show that sufficiency holds, imagine the space S' of outcomes of the throw of a fair, n-sided die to be partitioned into disjoint subsets X_i whose probabilities approximated those of the members of S under $P(\ )$. We can make the two sequences of probabilities as close as desired by making n sufficiently large. Let $P'(\ )$ be a uniform measure over the space of outcomes of tossing the die. $P'(\ )$ assigns the same probabilities to each partition element X_i as $P(\ )$ does to the corresponding element of S. Thus by condition (2), we need to show that $U(P', C')(H' \mid A') = P'(H' \mid A')$, where we take a primed proposition B' in S' to be the union of the partition elements of S' corresponding to the elements of the proposition B in S. This result follows since $U(P', C')$ will assign equal probabilities to the elements of A'. For the points in the partition elements have the same prior probabilities and since the posterior probability depends only upon the constraint and the prior probabilities assigned to points, and not upon the identity of these points, the points in A' must end up with equal probabilities in the posterior. A subset H of A (in S) corresponds to the union H' of partition elements of S' that are subsets of A'. Since all elements of A' have equal probabilities, we must have $U(P', C')(H' \mid A') = P'(H' \mid A')$.

7. CONDITIONALIZATION AND MAXENT

Jeffrey (1988) mentions a result of Diaconis and Zabell concerning when updating via Jeffrey's probability kinematics (a special case of MAXENT updating) can be represented by conditioning in an enriched space. Their result implies that this is always possible for a finite space. However, the conditioning statement need not be thought of in the way Skyrms suggests, as a statement in a higher-level space about probabilities in a lower-level space. Instead, it may be thought of as a statement that expresses the tacit evidence upon which the update is based. To use an example from Jeffrey (1965), if the update consists of new probabilities for the possible colors of a cloth observed by candlelight, the conditioning proposition would express the qualitative aspects of the experience that lead to the new probabilities. Of course, as Jeffrey (1988) has emphasized, the whole point behind his probability kinematics was to allow updating in cases in which such conditioning propositions are not available. It may be impossible for the believer to formulate a proposition that captures the relevant content of her experience.

We could, though, view the conditioning proposition X as having a purely *formal* role: it stands for a proposition that we could grasp had we adequate expressive resources and which stands in the same probabilistic relations to the expressible propositions as would this hypothetical proposition. Thus proposition X acts as the value of what Judea Pearl (1988) has called a 'dummy variable'. Pearl has led the way in developing data structures and algorithms suitable for efficient computer implementation of Bayesian updating. He introduced dummy variables precisely as a way of updating on uncertain evidence.

The question then arises as to whether or not this technique (call it DVU for 'dummy variable updating') might be a way of avoiding MAXENT in favor of conditionalization. The burden of the rest of this section is to show that it is not.

When the update consists of a new probability for a single proposition, DVU gives exactly the same result as MAXENT. However, if two or more propositions are simultaneously to be updated, the results may differ. It is instructive to see why this is so.

Let the probability measure $P(\ )$, shown in Table 1, be the prior assignment over the joint space of propositions A and B:

TABLE I

	P	Q
$A \& B$	0.05	0.332
$A \& -B$	0.45	0.332
$-A \& B$	0.45	0.332
$-A \& -B$	0.05	0.004

Suppose the constraints are that $P_{\text{new}}(A) = P_{\text{new}}(B) = 0.9$. DVU updates by introducing propositions X and Y such that $P(A \mid X) = 0.9$ and $P(B \mid Y) = 0.9$. DVU assumes that X is independent of B given A and Y is independent of A given B – i.e., that $P(X \mid A, B) = P(X \mid A)$ and $P(Y \mid A, B) = P(Y \mid B)$. Taking the posterior to be given by $Q(\ldots) = P(\ldots \mid X, Y)$, it is straightforward to calculate that $Q(\ )$ is as shown in Table 1. But this is not the posterior produced by MAXENT. $Q(A) = 0.664$, not 0.9. Thus $Q(\ )$ does not satisfy the given constraints, whereas MAXENT will always satisfy the constraints, provided they are consistent.

This does not mean that the update produced by DVU is wrong. If the new probabilities are based purely on two distinct observations or sensory experiences (i.e., they are what Skyrms (1985) calls 'observational probabilities'), then DVU yields the correct answer and MAXENT does not. For example, suppose you see an animal that for all the world looks like a duck, but at the same time hear emanating from it what sounds like intelligent speech. It seems to be delivering a discourse on Russell's theory of types, let's say. Based simply on its appearance *alone*, you might assign a 0.9 probability to its being a duck; and based upon its auditory emissions *alone*, you might assign a 0.9 probability to its understanding Russell's theory of types. But you would probably not end up with a 0.9 probability for its being a duck, all things considered. DVU seems to give the right answer in this case.

However, this does not mean that MAXENT is superfluous when the probabilities are observational probabilities. Separate out two things that happen when DVU is used. First, conditional constraints involving the dummy propositions are enforced; second, the dummy propositions are conditionalized upon. But in enforcing the conditional constraints, DVU *assumes* that each dummy proposition is independent of all other propositions conditional on the proposition for which it is evidence. Although this is a natural assumption in the stated epistemological setting, I argue that it can be rigorously justified only by appealing to MAXENT. On the view I am suggesting, DVU can be seen as a process of applying MAXENT to conditional constraints involving the dummy propositions and then conditioning on those propositions.

Consider, then, what happens when we start with a prior over propositions A and B and add constraints on the conditional probabilities $P(A \mid X)$ and $P(B \mid Y)$, where X and Y are the new dummy propositions. Since adding the constraints on these conditional probabilities should not cause a change in the prior over A and B, we assume that MAXENT, in addition to enforcing the new constraints, will also enforce constraints specifying that the marginal probabilities over the joint space of A and B have their prior values. Starting with a uniform prior and applying Csiszar's (1975) technique of iteratively satisfying the constraints, it is easy to show that MAXENT will produce a posterior in which X is independent of B given A and Y is independent of A given B. For if the constraint being satisfied is a constraint on the joint for A and B, these conditional independencies still hold, since probabilities conditional on an updated proposition remain the same. If the constraint is on $P(A \mid X)$ or on $P(B \mid Y)$, the technique of Lagrangian multipliers can be used to show that the conditional independencies are still satisfied. Hence they will be satisfied in the MAXENT posterior.

To make clear the argument of this section, what I have shown is that MAXENT can produce the same result as DVU when the constraints are interpreted as giving observational probabilities for certain propositions. Moreover, I argue that DVU makes certain assumptions about conditional independencies involving the dummy propositions, assumptions that MAXENT does not need to make, since they are automatically satisfied by MAXENT. In order to argue that MAXENT is superfluous in the case of observational probabilities, one would have to come up with an alternative justification of the conditional independence assumptions. No such justification has, to my knowledge, been presented. It does not seem possible, then, to replace MAXENT with an updating scheme based solely upon conditionalization.

Torrance, U.S.A.

NOTES

* I am indebted to Brian Skyrms for helpful discussions and correspondence on the topic of this paper.

[1] When the prior measure is uniform, minimizing relative entropy subject to certain constraints yields the same posterior as does maximizing entropy simpliciter subject to the same constraints. As does Skyrms, I intend the MAXENT principle to apply even when the prior is not uniform, although the terminology is somewhat misleading in that case.

REFERENCES

Csiszar, I.: 1975, 'I-divergence Geometry of Probability Distributions and Minimization Problems', *Annals of Probability* **3**, 146—158.

Diaconis, P. and Zabell, S.: 1982, 'Updating Subjective Probability', *Journal of the American Statistical Association* **77**, 822—830.

Domotor, Z.: 1980, 'Probability Kinematics and the Representation of Belief Change', *Philosophy of Science* **47**, 384—403.

Jaynes, E.: 1963, 'Brandeis Lectures', reprinted in Rosenkrantz (1983).

Jaynes, E.: 1978, 'Where Do We Stand on Maximum entropy?', in: R. D. Levine and M. Tribus (eds.), *The Maximum Entropy Formalism*, M.I.T. Press, Cambridge, Mass. Reprinted in Rosenkrantz (1983).

Jeffrey, R.: 1965, *The Logic of Decision*, McGraw Hill, N.Y. (2nd ed., University of Chicago Press, Chicago, 1983).

Jeffrey, R.: 1988, 'Conditioning, Kinematics, and Exchangeability', in B. Skyrms and W.

L. Harper, (eds.), *Causation, Chance, and Credence*, Kluwer Academic Publishers, Dordrecht.
Pearl, J.: 1988, *Probabilistic Reasoning in Intelligent Systems*, Morgan Kaufmann Publishers, Inc., San Mateo.
Rosenkrantz, R. D.(ed.) : 1983, *E. T. Jaynes: Papers on Probability, Statistics, and Statistical Physics*, D. Reidel Publishing Co., Dordrecht.
Shore, J. and Johnson, R.: 1980, 'Axiomatic Derivation of the Principle of Maximum Entropy and the Principle of Minimum Cross-entropy', *IEEE Transactions in Information Theory* IT-26 **1**, 26—37.
Skyrms, B.: 1980, 'Higher Order Degrees of Belief' in D. H. Mellor (ed.), *Prospects for Pragmatism*, Cambridge University Press, Cambridge.
Skyrms, B.: 1985, 'Maximum Entropy Inference as a Special Case of Conditionalization', *Synthese* **63**, 55—74.
Skyrms, B.: 1987, 'Updating, Supposing, and Maxent', *Theory and Decision*, 225—246.
Van Fraassen, B.: 1980, 'Rational Belief and Probability Kinematics', *Philosophy of Science* **47**, 165—187.
Williams, P. M.: 1980, 'Bayesian Conditionalization and the Principle of Minimum Information', *British Journal for the Philosophy of Science* **32**, 375—379.

WOLFGANG LENZEN

LEIBNIZ ON ENS AND EXISTENCE

HOMAGE

The issue of existence has been one of Karel Lambert's main philosophical concerns for more than two decades. It constitutes the central motivating factor not only for his pioneering development of free logic but also for his less formal philosophical works.[1] Moreover it was his Gastvortrag on existence once given at the Department of Philosophy of the University of Regensburg, West Germany, in the early 70ies that marks the beginning of our friendship, although, as Joe never failed to mention when introducing me to one of his colleagues, a critical remark of mine 'destroyed' his then theory of inexistent, fictional objects. Therefore it will certainly not be inappropriate if my contribution to this Festschrift centers around the very issue of existence.

There is a twofold reason, however, why I shall not simply state my own views on existence here. First, my favourite way of construing existence within the framework of possible worlds semantics is neither very exciting nor very original. It basically follows the lines of Dana Scott's *Advice on Modal Logic*,[2] and it has been presented at some length elsewhere.[3] Second, Gottfried Wilhelm Leibniz's views on existence are not only more exciting and much more original than mine, but they also reveal the intimate connection between his metaphysical and logical ideas. Leibniz himself was well aware of this connection, and he never became tired of stressing "... that the true Metaphysics is hardly different from the true Logic" (GP 4, 292).[4] The aim of this paper is to provide a consistent reconstruction of a small part of his 'true Metaphysics' in terms of his 'true Logic'.

THE LOGIC OF CONCEPTS

Leibniz's logic is a second order logic of concepts based upon a sentential calculus of strict implication. Somewhat more exactly: it is a quantificational extension of the algebra of concepts, which itself constitutes but an 'intensional' counterpart of the ordinary, 'extensional'

Wolfgang Spohn et al. (eds.), Existence and Explanation, 59—75.

algebra of sets. The quantifiers in question first range over concepts, but quantification over individuals can be modelled by quantifying over individual-concepts which in turn may be introduced by a purely logical definition. In some more detail:

The syntax of the algebra of concepts, L1, consists of:

- a denumerable set of concept letters $A, B, C, \ldots$;
- the operators of conceptual negation: $\sim A$ — 'Not-A' —, and conceptual conjunction: AB — 'A and B';
- the relation of conceptual containment: $A\mathrm{c}B$ — 'A contains B';
- the propositional operators of negation: $\neg\alpha$, conjunction: $\alpha \,\&\, \beta$, and logical implication: $\alpha \rightarrow \beta$.

The language of the universal calculus, L2, contains in addition:

- a denumerable set of concept-variables $X, Y, Z, \ldots$; and
- the conceptual quantifier: $\exists$ — 'for at least one (concept)'.

Terms and sentences are determined as usual, and some further operators can be introduced by definition:

Def. 1

$P(A) =_{\mathrm{df}} \neg\exists Y(A\mathrm{c}Y{\sim}Y)$ — '(Concept) A is possible (self-consistent)'[5]

$A = B =_{\mathrm{df}} (A\mathrm{c}B) \,\&\, (B\mathrm{c}A)$ — 'A coincides with B'

$0 =_{\mathrm{df}} {\sim}(A{\sim}A)$ — the empty (tautological) concept

$\forall X\alpha =_{\mathrm{df}} \neg\exists X\neg\alpha$ — 'for every (concept) X: α'

When Leibniz began to develop the algebra of concepts sometime around 1679, he tacitly presupposed certain laws of propositional logic. After he had found a satisfactory version of L1, he was struck by the idea that propositional logic might be derived from concept logic simply by considering all propositions 'instar terminorum' and by holding that:

> ... whatever is said of a term which contains another term can also be said of a proposition from which another proposition follows. ... [T]hat a proposition follows from a proposition is simply that a consequent is contained in an antecedent, as a term in a term. By this method we reduce inferences to propositions, and propositions to terms. (P, 85, 87)

If one carries out this idea[6] the algebra of concepts gets transformed into an algebra of propositions which constitutes a full-blown calculus of strict implication. In what follows, however, the propositional basis of the system of Leibniz's logic will be left out of consideration.

Therefore the 6 subsequent principles (listed together with the original formulation from the GI) already form a complete axiomatization[7] of L1:

Ax1 AcA – 'B is B' – §37

Ax2 AcB & $BcC \rightarrow AcC$ – 'if A is B and B is C, A will be C' §19

Ax3 AcB & $AcC \leftrightarrow AcBC$ – 'That A contains B and A contains C is the same as that A contains BC' – §35

Ax4 $\sim\sim A = A$ – 'Not-not $A = A$', – §96

Ax5 $\neg P(A\sim B) \leftrightarrow AcB$ – '... A Not-B is impossible, i.e. A contains B', – §128[a]

Ax6 AcB & $P(A) \rightarrow P(B)$ – 'If A contains B and A is true B also is true.'[9] – §55

I want to refrain from stating axioms also for the stronger calculus L2. Some typical Leibnitian laws for the quantifiers will be mentioned below, and some more such principles have been discussed elsewhere.[10] As far as I know, however, it is still an open problem how a complete axiomatization of L2 would look like.

Neglecting the sentential part of the universal calculus, the semantics can be defined in accordance with Leibniz's ideas as follows:

Def. 2
A model of the logic of concepts is a function V over a non-empty domain D (of possible objects) which assigns to each term A an extension $V(A) \subseteq D$ and to each proposition α a truth-value $V(\alpha) \in \{0, 1\}$ such that:

(1) $V(\sim A) = \overline{V(A)}$;

(2) $V(AB) = V(A) \cap V(B)$;

(3) $V(AcB) = 1$ iff $V(A) \subseteq V(B)$;

(4) $V(\exists X\alpha) = 1$ iff for some interpretation V^* such that $V^*_A = V$ (i.e., such that $V^*(B) = V(B)$ for every concept letter B except perhaps for $B = A$) $V^*(\alpha[A]) = 1$, where A must not appear in $\exists X\alpha$.

The corresponding conditions for the defined operators amount to:

(5) $V(P(A)) = 1$ iff $V(A) \neq \emptyset$;

(6) $V(A = B) = 1$ iff $V(A) = V(B)$;

(7) $V(0) = D$;

(8) $V(\forall X\alpha) = 1$ iff $V^*(\alpha[A]) = 1$ for every interpretation V^* such that $V^*_A = V$ (here, again, A must not occur in $\forall X\alpha$).

As usual a sentence α will be called logically true iff α is satisfied by every model V. Each model V represents an extensional interpretation of 'intensional' concept logic. According to clause (1), the extension of the negative concept $\sim A$ is just the complement of the extension of A, i.e. the set of individuals that fail to fall under the concept A; according to (2) the extension of the conjunction AB is the intersection of the two extensions, i.e. the set of individuals that fall under both concepts; according to (3), concept A contains concept B just in case the extension of A is contained in the extension of B. This requirement reflects the so-called law of reciprocity of extension and intension:

The method of concepts is contrary to the method of individuals, namely: If all the men are part of all the animals, or if all men are among all animals, vice versa the concept of animal will be in the concept of man; and such as there are more animals beside the men, so something has to be added to the idea of animal in order to obtain the idea of man. For by increasing the conditions, the number decreases.[11]

The law of reciprocity entails that whenever two concepts A and B have exactly the same extension, then their intensions are mutually contained in each other: AcB & BcA, and hence by Def. 1 they coincide: $A = B$. This is also reflected by clause (6) of Def. 2. Now it might be objected that there are intensionally different concepts which nevertheless happen to have exactly the same extension, e.g. the concepts 'animal with a heart' and 'animal with a kidney'. To this Leibniz would reply that the extensions must be taken from the set of all possible individuals and not only from the set of actually existing ones. Since there are possible animals which have, e.g., a heart but no kidney, it then follows that not only the intensions but also the extensions of both concepts do indeed differ from one another. The same observation also helps to remove the prima facie implausibility of clause (5) according to which concept A is possible iff A has a non-empty extension. Although there certainly are many self-consistent concepts which have an empty extension with regard to the set of actually existing individuals (e.g., the concept of a unicorn), their relevant extension with regard to the set of all possible individuals nevertheless will be non-null. The remaining clauses 7, 4, and 8 are straightforward: as a consequence of the law of reciprocity, the tautological concept 0 which has minimal intension hence has maximal extension, *viz*. the entire domain D; furthermore, an existentially (or universally) quantified statement α is true iff it is true for at least one (or for every) concept A.

EXISTENCE OF INDIVIDUALS

Let us now consider §§71—72 GI (P, 65):

(71) What is to be said about the proposition '*A* is an existent' or '*A* exists'? Thus, if I say about an existing thing, '*A* is *B*' it is the same as if I were to say '*AB* is an existent'; e.g. 'Peter is a denier', i.e. 'Peter denying is an existent'. The question here is how one is to proceed in analysing this; i.e. whether the term 'Peter denying' involves existence, or whether 'Peter existent' involves denial — or whether 'Peter' involves both existence and denial, as if you were to say 'Peter is an actual denier', i.e. is an existent denier; which is certainly true. Undoubtedly, one must speak in this way; and this is the difference between an individual or complete term and another. For if I say 'Some man is a denier', 'man' does not contain 'denial', as it is an incomplete term, nor does 'man' contain all that can be said of that of which it can itself be said.

(72) So if we have *BY*, and the indefinite term *Y* is superfluous (i.e., in the way that 'a certain Alexander the Great' and 'Alexander the Great' are the same), then *B* is an individual. If there is a term *BA* and *B* is an individual, *A* will be superfluous; or if $BA = C$, then $B = C$.

Let us begin by clarifying the central notions 'existing', 'individual', and 'individual-term'. Leibniz has often been blamed for not carefully distinguishing between terms and their denotations.[12] The quoted passage certainly justifies such a criticism, but Leibniz's rather careless use of the word 'individual' to refer to individual-terms does not give rise to serious misunderstandings. One may assume that there is a one-to-one-correspondence between individuals and individual-terms, and the context makes perfectly clear what Leibniz is talking about. What has to be kept in mind, however, is that an individual-term for Leibniz nevertheless is a concept, i.e. an 'intensional' entity which may contain (or be contained in) other 'intensional' concepts. Hence its extension must be conceived of as a subset of D and not, as is the case with an individual-term, i.e. a name, of ordinary predicate logic, as an element of D. E.g., the extension of 'Petrus' is not the individual Petrus but the unit-set containing exactly that individual. With '$I(A)$' abbreviating 'A is an individual-concept', we thus may continue **Def. 2**:

(9) $V(I(A)) = 1$ iff there is some individual $a \in D$ such that $V(A) = \{a\}$;

As regards the notion of existence, Leibniz is treating it on a par with the other concepts by forming corresponding conjunctions 'Petrus existens', 'abnegans existens' which enter into the fundamental relation 'c'. Therefore 'existens' may be abbreviated by a distinguished concept letter, say E, to be interpreted extensionally, like any other concept letter, as a certain subset $V(E) \subseteq D$. According to Leibniz, the extreme cases that $V(E) = \emptyset$ or that $V(E) = D$ should be excluded. For he not only believed it to be 'impossible that nothing exists' (AV, 67), but he also held the view that not all of the possible individuals are compossible and that therefore some individuals will not be created by God but will remain mere possibles.[13] We want to leave this issue undecided, however, and only postulate:

Def. 2 continued
(10) $V(E) \subseteq D$;

Now, generalizing from the above examples, Leibniz is maintaining that whenever A is the complete term of an existing individual, then the statement 'A is B' is equivalent both to (i) 'AB is Existing' and to (ii) 'A Existing is B', and also to (iii) 'A is Existing B':

Th1 $I(A) \,\&\, A\text{c}E \rightarrow (A\text{c}B \leftrightarrow AE\text{c}B)$

Th2 $I(A) \,\&\, A\text{c}E \rightarrow (A\text{c}B \leftrightarrow AB\text{c}E)$

Th3 $I(A) \,\&\, A\text{c}E \rightarrow (A\text{c}B \leftrightarrow A\text{c}BE)$

These principles turn out to be theorems of the algebra of concepts. As regards **Th1**, observe that:

Th4 $AB\text{c}A$ — 'AB est A' (C, 263)

Th5 $AB\text{c}B$ — 'AB is B' (GI, §38)

are corollaries of the conjunction axiom **Ax3**. Hence in particular $AE\text{c}A$, so that $A\text{c}B$ entails by the transitivity of 'c' that $AE\text{c}B$. Furthermore, the basic relation of conceptual containment is reducible to a condition in terms of conjunction and coincidence:

Th6 $AcB \leftrightarrow A = AB$ – "Generally, '*A* is *B*' is the same as '*A* = *AB*'" (GI, §83).

Hence premise AcE of **Th1**, i.e. $A = AE$, together with the assumption $AEcB$ conversely entails AcB by the substitutivity of identicals. A proof of **Th2** is obtained in a strictly analogous way. With respect to **Th3**, AcE in conjunction with AcB immediately entails (**Ax3**) that $AcBE$; and the converse inference, $AcBE \rightarrow AcB$, is entirely trivial.

In the preceding demonstrations no use was made of the specific character of individual concepts, i.e. of their completeness. What, then, had Leibniz in mind when he went on to explain: "Undoubtedly, one must speak in this way; and this is the difference between an individual or complete term and another." My answer will be surprising at first sight: the difference between an individual concept and an 'ordinary' one is that the proposition '*A* exists' or '*A* est existens' may only in the former but not in the latter case be regarded as a relation of conceptual containment and hence be formalized as 'AcE'. Why this is the case shall not be explained before the end of the next section, when §§144–150 GI will be under discussion. Here I first want to deal with some other criteria for distinguishing individual-concepts from 'ordinary' concepts.

A first difference between an incomplete concept as, e.g., 'man', and a complete one as, e.g., 'Alexander the Great' is indicated by saying that 'man' does not "contain all that can be said of that [e.g., Alexander the Great] of which it ['man'] can itself be said". Thus 'Alexander the Great is an emperor' is true, but 'emperor' cannot be generally said of (or attributed to) 'man', since 'Man is emperor' is always[14] to be understood as the universal proposition that every man is an emperor. This observation illustrates the 'incompleteness' of an 'ordinary' concept, but it does not yet provide a clear definition of the completeness of individual-concepts. For to say that an individual-concept *B* "contains all that can be said of that of which it [*B*] can be said" seems only to express the triviality that *B* contains every concept *Y* which it contains.

A second difference between individual concepts and 'ordinary' concepts is vaguely outlined in the same sentence when Leibniz remarks that from the truth of the particular proposition 'Some man is a denier' it does not follow that the universal proposition 'Every man is a denier' or, for short 'Man is denier' be true as well ("'man' does not contain 'denial'"). Here one evidently has to add the unspoken claim that the corresponding inference from a particular to a universal proposition does hold if the subject term is an individual-concept. This stands in close connection with the parenthetical remark of §72: "'a certain Alexander the Great' and 'Alexander the Great' are the same", and also with the following passage from another fragment:

> How is it that opposition is valid in the case of singular propositions — e.g. 'The Apostle Peter is a soldier' and 'The Apostle Peter is not a soldier' — since elsewhere a universal affirmative and a particular negative are opposed? Should we say that a singular proposition is equivalent to a particular and to a universal proposition? Yes, we should. . . . For 'some Apostle Peter' and 'every Apostle Peter' coincide, since the term is singular. (P, 115)

Let us see if this 'Wild Quantity Thesis'[15] — WQT, for short — can be verified within the universal calculus of concepts. Observe, first, that a universal affirmative proposition is not only representable, in L1, by the formula '$A\mathrm{c}B$', but also, in L2, by '$\forall Y(YA\mathrm{c}B)$'. On the one hand, $\forall Y(YA\mathrm{c}B)$ immediately entails for $Y = A$ that $AA\mathrm{c}B$ and thus, because of the trivial law

Th7 $A = AA$ — '*A*, *AA*, *AAA*, etc. coincide' (GI, §18),

also $A\mathrm{c}B$; conversely $YA\mathrm{c}B$ follows, for arbitrary Y, from the assumption $A\mathrm{c}B$ and from **Th5** by the transitivity of 'c'. Hence:

Th8 $A\mathrm{c}B \leftrightarrow \forall Y(YA\mathrm{c}B)$.

In the case of a singular proposition $A\mathrm{c}B$, **Th8** may be interpreted as saying that the subject term A, e.g., 'Apostle Peter', is equivalent to the universally quantified term YA, 'every Apostle Peter', as maintained in WQT. Similarly, Leibniz tried to represent the particular proposition 'Some A is B' as '$YA\mathrm{c}B$', i.e. less elliptically as '$\exists Y(YA\mathrm{c}B)$'.[16] This formalization faces a minor difficulty, however. For Leibniz himself noted in §49 GI that any proposition $AC\mathrm{c}B$ entails, by way of existential generalization, $\exists Y(AY\mathrm{c}B)$: "If AB is C, it follows that AY is C; or, it follows that some A is C. For it can be assumed, by 23,[17] that $B = Y$." Thus in view of $AB\mathrm{c}B$ (**Th5**), it is an immediate truth of L2 that for at least one concept Y: $AY\mathrm{c}B$. Hence the latter formula cannot adequately express the content of the particular affirmative proposition.

As Leibniz recognized in a later fragment,[18] this difficulty may be overcome by demanding more exactly that for some Y compatible with A AY contains B. The second part of WQT thus can be rephrased as stating that — for any individual-concept A — the sentences AcB and $\exists Y(P(AY) \,\&\, AYcB)$ are equivalent:

Th9 $\quad I(A) \rightarrow (AcB \leftrightarrow \exists Y(P(AY) \,\&\, AYcB))$.

To show the validity of **Th9**, we will first have to develop a logical definition of the 'completeness'-condition for individual concepts. In the concluding sentence of §72, Leibniz defines concept A to be 'superfluous' (with respect to concept B) iff (for every C) $BA = C$ entails that $B = C$ entails that $B = C$. This condition may be simplified by just requiring that $BA = B$, or, in view of **Th6**, BcA. For, on the one hand, substituting 'BA' for 'C' yields that $BA = BA$ entails $B = BA$; conversely, if $BA = B$ then (for any C) $BA = C$ entails that $B = C$. Hence A is superfluous with respect to B just in case that A is already contained in B. Now, when Leibniz goes on saying "If there is a term BA and B is an individual, A will be superfluous", he seems to maintain that any term A is superfluous with respect to any individual term B. But this is absurd since otherwise an individual-concept B would be 'completely complete' in the sense of containing every concept A, in particular besides A also $\sim A$, and hence B would be inconsistent.

To resolve this difficulty, observe that Leibniz begins the sentence in question by saying "Si sit terminus BA" which Parkinson renders as "If there is a term BA". In other contexts, this translation surely would be appropriate to express the sense of a mere stipulation ("Let there be a term BA . . ."); in the present context, however, I would suggest, Leibniz meant to say: "Let the term BA *be*", i.e., *be a consistent term*, i.e. let us suppose that $P(BA)$! There are several other passages within and without the GI where Leibniz paraphrases the condition of self-consistency of a concept A just by saying: 'A is', e.g. in §144 which will be analyzed in the next section. Therefore the interpretation of "Si sit terminus BA" as meaning 'Let BA be a possible term' is not implausible, and it entails the necessary condition: B is an individual-concept only if — unlike other concepts — B is complete in the precise sense of already containing any concept A with which it is compatible (i.e. for which $P(BA)$ holds). Since A here stands for any arbitrary concept, it may be replaced by a concept-variable Y and then be bound by a universal quantifier: $I(B) \rightarrow \forall Y(P(BY) \rightarrow BcY)$. That this is what Leibniz had in mind is evidenced by the fact that the converse implication $\forall Y(P(BY) \rightarrow BcY) \rightarrow I(B)$ is recognized by him as a sufficient condition for a concept B to be an individual-concept when he says: "So if BY is [possible], and the arbitrary[19] indefinite term Y is superfluous, then B is an individual." If one adds the trivial condition $P(B)$, then the following Leibnitian definition of individual-concepts is obtained:

Def.3
$I(A) =_{df} P(A) \& \forall Y(P(AY) \rightarrow AcY)$[20]

This definition is semantically adequate. It is easy to verify that for any extensional model V: $V(I(A)) = 1$ iff there is an individaul $a \in D$ such that $V(A) = \{a\}$. Moreover, Def. 3 enables us to prove the open part of WQT, i.e. **Th9**, as follows: If $I(A)$ and AcB, then — trivially — $AAcB$ and $P(AA)$, from which $\exists Y(P(AY) \& AYcB)$ follows by existential generalization; conversely, let there be some Y such that $P(AY)$ & $AYcB$; since A is presupposed to be an individual-concept, $P(AY)$ according to Def. 3 implies that AcY, i.e. $A = AY$, so that $AYcB$ yields the desired AcB.

'ESSENTIAL' AND 'EXISTENTIAL' PROPOSITIONS

So far we have been concerned with the truth-conditions for attributing existence to individuals. Let us now consider §§144—150 GI where Leibniz investigates the truth-conditions for corresponding non-singular categorical propositions.

Propositions are either essential or existential, and both are either *secundi adjecti* or *tertii adjecti*. An essential proposition *tertii adjecti* is, for example, '*A* circle is a plane figure'. An essential proposition *secundi adjecti* is, for example, '*A* plane figure having a constant relation to some one point is' [est].[21] I say 'is', i.e. 'it can be understood', 'it can be conceived', among various figures there is one which also has this nature, just if I were to say '*A* plane figure having a constant relation to some one point is an entity or a thing' [est ens sive res]. An existential proposition *tertii adjecti* is 'Every man exists liable to sin'. This is manifestly an existential or contingent proposition. An existential proposition *secundi adjecti* is '*A* man liable to sin exists, i.e. is an actual entity' [existit seu est ens actu].

(145) From every proposition *tertii adjecti* a proposition *secundi adjecti* can be made, if the predicate is compounded with the subject into one term and this is said to be or to exist [esse vel existere], i.e. is said to be either a thing howsoever or an actually existing one [esse res sive utcunque, sive actu existens].

(146) The particular affirmative proposition, 'Some A is B', transformed into a

proposition *secundi adjecti*, will be '*AB* is' [est], i.e. '*AB* is a thing', either possible or actual, depending on whether the proposition is essential or existential. . . .

(148) The particular negative proposition, 'Some *A* is not *B*', will be transformed into a proposition *secundi adjecti* as follows: '*A* not-*B* is', i.e. '*A* which is not *B* is a certain thing'; possible or actual, depending on whether the proposition is essential or existential.

(149) The universal negative proposition is transformed into a proposition *secundi adjecti* by the negation of the particular affirmative proposition. E.g., 'No *A* is *B*', i.e. '*AB* is not' [non est], i.e. 'AB is not a thing' . . .

(150) The universal affirmative proposition is transformed into a proposition *secundi adjecti* by the negation of the particular negative proposition, so that 'Every *A* is *B*' is the same as '*A* not-*B* is not, i.e. is not a thing' [non est, seu non est res] . . .

The omissions in §§146, 149, and 150 as well as the entire §147 pertain to an alternative reduction of universal propositions which, however, may be neglected since Leibniz himself recognized it as 'less apt'. If it is observed that the 'being' of a concept *A* is tantamount to its being possible, the 'essential' interpretation of the categorical forms can be formalized in the following:

Schema 1

U.A.	$\neg P(A \sim B)$	$\neg P(AB)$	U.N.
P.A.	$P(AB)$	$P(A \sim B)$	P.N.

On the other hand, the U.A. is usually represented in L1 simply by 'AcB': "(47) '*A* contains *B*' is the universal affirmative proposition with respect to the subject *A*." In view of the medieval law of conversion,[22] the U.N. can be formalized as '$Ac \sim B$', and by the laws of opposition the negations of these formulas represent the P.A. and P.N.:

Schema 2[23]

U.A.	AcB	$Ac \sim B$	U.N.
P.A.	$\neg(Ac \sim B)$	$\neg(AcB)$	P.N.

Leibniz's thesis of the reducibility of the categorical forms *tertii adjecti* to propositions *secundi adjecti* may thus be restated as the claim that Schema 1 is provably equivalent to Schema 2. This, however, easily follows from **Ax5** and from the 'homogeneity'[24] of both schemata.

Let us now turn to the 'existential' interpretation of the categorical forms. Just as the truth of the 'essential' P.A. "*AB* est res . . . possibilis" according to our semantics requires that there is at least one possible individual $x \in D$ such that x is an AB, i.e. x is both an A and a B, so the stronger, 'existential' P.A. "*AB* est res nempe . . . actualis" should be considered as true iff there is an actually existing individual x which is both an A and a B. How can this be expressed, however, within the logic of concepts?

There is one fragment (GP, 7, 211—7) where Leibniz attempted to understand 'ens' as a concept letter and where accordingly he interpreted 'A est Ens' as having the logical form 'AcEns'. *If* this were admissable, then one might transform '*AB* est Ens' by means of **Th6** into $AB = AB$Ens as Leibniz did in fact try at P, 117: "Similarly, 'Some A is B', that is, '*AB* is an entity' [*AB* est Ens], can also be expressed as '*AB* and *AB*-entity [*AB* Ens] are equivalent'." In the same way, one might be seduced to formalize the 'existential' P.A. '*AB* est existens' by means of the concept E as 'ABcE', or, '$AB = ABE$'. This approach, however, is easily seen to lead to inconsistencies. For example, the assumption that a certain concept A is either 'ens' or 'existens' — if formally represented as 'Ac*Ens*' or 'AcE(xistens)' — would entail by the trivial **Th4** and **Ax2** that for any B 'ABc*Ens*' or 'ABcE(xistens)', hence in particular '$A \sim A$c*Ens*' or '$A \sim A$cE(xistens)' in flat contradiction to the fact that $A \sim A$ is inconsistent, i.e.: "A not-A is not a thing" (GI §171; P, 83), neither a possible nor an actually existing one.

Fortunately, in another fragment (C, 270—3) Leibniz discovered the way out of this problem when asking himself:

> . . . if the way of transforming logical propositions into terms by adding just 'ens' or 'non ens' also works in the case of existential propositions. . . . For example: 'Some pious [man] is poor', i.e. 'Pious poor is existing'. . . . Let us see whether 'existing' can also be moved into the term so that only 'ens' or 'non Ens' remains. Such that 'Pious poor is existing' yields 'Pious poor existing is Ens'. (C, 271)

Generalizing from this example, an 'existential' P.A. '*AB* is existing' shall be reduced to a proposition *secundi adjecti* by maintaining the conjunction ABE(xistens) to be 'Ens', i.e. to be possible or self-consistent: $P(ABE)$! Similarly, an 'existential' P.N. 'A Not-B is existing' will have to be represented by $P(A \sim BE)$, as Leibniz illustrates when transforming "quidam pius non est pauper, seu Pius non pauper est existens" into "Pius existens non pauper est Ens seu possibile" (ibid.). Finally, 'existential' versions of the universal propositions can be obtained by negating P.A. and P.N., respectively, e.g.: "Pius existens non tribulatus est non ens, seu impossibile"; "Justus derelictus existens est non Ens, seu impossibile" (ibid.). In sum, then, one arrives at the following 'existential' counterpart of Schema 1:

Schema 1_E

U.A.$_{\cdot E}$	$\neg P(A \sim BE)$	$\neg P(ABE)$	U.N.$_{\cdot E}$
P.A.$_{\cdot E}$	$P(ABE)$	$P(A \sim BE)$	P.N.$_{\cdot E}$

This formalisation correctly represents the 'existential' interpretation of the categorical propositions. E.g., it follows from Def. 2 that $V(P(ABE)) = 1$ iff $V(ABE) \neq \emptyset$, i.e. iff $V(AB) \cap V(E) \neq \emptyset$, i.e. iff there is an actually existing individual x ($\in V(E)$!) which is both an A and a B! Similarly, the 'existentially' interpreted U.A. turns out to be true just in case that every actually existing individual x, which is an A, also is a B. And the same holds *mutatis mutandis* also for the negative propositions. To conclude the discussion of §§144—150, let me resolve the following, perplexing asymmetry between 'ens' and 'existens'.

While the mere 'being' of a concept A so far has always been treated in the form '$P(A)$', A's 'existing' was considered as a genuine attribution of the the form '$A\text{c}E$'. Does this mean that only 'existens' but not 'ens' is a real concept? No! The apparent asymmetry does not result from any logical difference between the two modes of being, but simply from a different logical behavior of individual-concepts and of 'ordinary' concepts. If A is an individual-concept, then '$A\text{c}E$' correctly expresses that the individual denoted by 'A', say, a, is an existing one. Similarly, one might well express the weaker condition that a is a possible individual in the same logical form '$A\text{c}$ . . .', namely by filling into the lacuna a very peculiar concept letter that has been ignored so far: the empty concept 0! In a way, however, this is redundant, since for any A:

Th10 $A\text{c}0$

is provable; hence in particular $I(A) \rightarrow A\text{c}0$ always holds. But this is what should be expected anyway, for, after all, every individual is a possible individual, so that, provably, its complete concept A must be self-consistent!

On the other hand, if A is an 'ordinary' concept, then not only its being 'ens' can be formalized in the form $P(A)$, but also its being 'existens' might be expressed in the same logical form . . . (A). Letting the new sentence '$R(A)$' abbreviate the condition that the concept A is realized in the sense that at least one actually existing individual x falls under the concept A:

Def. 2 continued
(11) $V(R(A)) = 1$ iff $V(A) \cap V(E) \neq \emptyset$

then the following principles which re-establish the symmetry between both kinds of being become provable:

Th11 $R(A) \leftrightarrow A \not\subset {\sim} E$

Th12 $P(A) \leftrightarrow A \not\subset {\sim} 0.$

Now, the completeness of an individual-term A is tantamount to the condition that for any Y: $A \mathrm{c} Y$ iff $A \not\subset {\sim} Y$. Therefore

Th13 $I(A) \rightarrow (R(A) \leftrightarrow A \mathrm{c} E)$[25]

holds which explains why the existence of an individual may be expressed by '$A\mathrm{c}E$'.

CONCLUSION

In this paper only a very small part of Leibniz's 'true Metaphysics' has been dealt with. As a matter of fact, not even the underlying 'true Logic' has been reconstructed at full length. Let me just sketch in barest outline how a more complete reconstruction of Leibniz's theory of 'ens' and 'existence' would have to proceed. First, quantification over individuals should be modelled by restricting the quantifiers to individual concepts: e.g., $\vee X\alpha =_{df} \exists X(I(X) \,\&\, \alpha)$. Second, the relation of compossibility of individuals should be defined to obtain iff X and Y will co-exist in some possible world, i.e. iff they possibly coexist: $CP(X, Y) =_{df} \Diamond(X\mathrm{c}E \,\&\, Y\mathrm{c}E)$. Third, possible worlds will be constructed as maximal sets of compossible individuals. Fourth, a new relativized notion of 'existence' in a possible world w, E^w, will be introduced such that $E^w(X)$ is true exactly for those individuals X which belong to w. Fifth, the actual world w^* may be singled out from the set W of all possible worlds by its containing the greatest possible number of elements. Sixth and lastly, our 'ordinary' notion of (actual) existence, E, will be equated with E^{w^*}. This chain of logical moves seems to stand behind Leibniz's insight that 'existens' "can be defined as 'that which is compatible with more things than anything else which is incompatible with it'." (P, 51)

In his recent monograph on *The Philosophy of Leibniz*, Benson Mates has rightly remarked that

... we cannot be content with merely paraphrasing his actual words. ... For in choosing among the various possible interpretations of an author's dicta, one has to consider logical relationships. Ceteris paribus, an interpretation is preferable if it preserves consistency and relevance. (o.c., p. 51)

I hope that the preceding exposition of Leibniz's theory of Ens and Existence satisfies the latter criteria and thus may find the approval both of my friend Karel Lambert and of my highly-esteemed colleague Benson Mates.

Department of Philosophy
University of Osnabrück
Osnabrück, Germany

NOTES

[1] Cf. especially 'On Logic and Existence', *Notre Dame Journal of Formal Logic* **6**, 1965, 135—141; 'Impossible Objects', *Inquiry* **17**, 1974, 303—314; 'On the Philosophical Foundations of Free Logic', *Inquiry* **24**, 1981, 147—203; and the monograph *Meinong and the Principle of Independence*, Cambridge University Press, [Cambridge] 1983.

[2] In Karel Lambert (ed.), *Philosophical Problems in Logic*, Reidel, Dordrecht, 1969.

[3] The idea is to let the universe of (epistemic) discourse just be some non-empty set of entities called 'possible individuals' whereas the so-called 'actually existing individuals' simply form a subset (not even necessarily a genuine one) of the former set. My reasons for following this strategy especially in the context of epistemic logic have been stated in 'Knowledge, Belief, Existence, and Quantifiers — A Note on Hintikka', *Grazer Philosophische Studien* **2** 1976, 55—65; cf. also section 8.2 of my book *Glauben, Wissen und Wahrscheinlichkeit*, Wien, 1980.

[4] The following abbreviations are used:

AV = *Vorausedition zur Reihe VI — Philosophische Schriften — in der Ausgabe der Akademie der Wissenschaften* ..., ed. by the Leibniz-Forschungsstelle, University of Münster, Münster, 1982ff.;

C = L. Couturat (ed.), *Opuscules et fragments inédits de Leibniz*, Paris, 1903;

GI = F. Schupp (ed.): *Generales Inquisitiones de Analysi Notionum et Veritatum*, Meiner Verlag, Hamburg; 1982;

GP = C. J. Gerhardt (ed.), *Die philosophischen Schriften von G. W. Leibniz*, 7 vol., Berlin, 1875—1890;

P = G. H. R. Parkinson (ed.), *Leibniz: Logical Papers*, Clarendon Press, Oxford, 1966.

[5] Cf., e.g., §1 GI: "*A* Not-*A* is a contradictory term. That which does not contain a contradictory term, i.e. *A* not-*A*, is possible. That which is not *Y* not-*Y* is possible" (P, 54).

[6] As is done in my 'Leibniz's Calculus of Strict Implication' in J. Srednicki (ed), *Initiatives in Logic*, Nijhoff, Dordrecht, 1987, pp. 1—35.

[7] A proof of completeness may be found in my 'Leibniz und die Boolesche Algebra', *Studia Leibnitiana* **16**, 1984, 187—203.

[8] Unfortunately, in Parkinson's translation this sentence which states that if the particular negative proposition is false, then *A* Not-*B* is impossible, i.e. [!] *A* contains *B*, has been omitted.

[9] The word 'true' seems to indicate that Leibniz here wants to express something like the rule of detachment; he goes on to explain, however: "By a false letter I understand either a false term (i.e. one which is impossible, or, is a non-entity) or a false proposition . . .". Thus, if the letters *A* and *B* are taken to stand for concepts, the quoted sentence does express **Ax6**, while the equally admissable interpretation of *A* and *B* as propositions leads to the rule of modus ponens.

[10] Cf. my 'Unbestimmte Begriffe' bei Leibniz', *Studia Leibnitiana* **16**, 1984, 1—26.

[11] C, 235; cf. also the famous passage from the *Nouveaux Essais* where Leibniz uses the expressions 'extension' and 'intension'; GP, 5, 469.

[12] E.g. in B. Mates, *The Philosophy of Leibniz*, Oxford University Press, Oxford, 1986.

[13] Cf., e.g., C, 530 and 534, #7.

[14] Except for the *Dissertatio de Arte Combinatoria* of 1666 and for one early fragment of around 1676 (C, 321—4), Leibniz always followed the convention of expressing the universal affirmative proposition 'Omne *A* est *B*' simply as '*A* est *B*', i.e. *AcB*. Thus, e.g., in the *Specimen Calculi universalis* of 1679 he declared "A universal affirmative proposition will be expressed here as follows: *a* is *b*, or, (every) man is an animal. We shall, therefore, always understand the sign of universality to be prefixed" (P, 33). He stuck to this convention throughout his later logical writings.

[15] Cf. Englebretsen, 'A Note on Leibniz's Wild Quantity Thesis', *Studia Leibnitiana* **20** 1988, 87—9.

[16] Cf. §48 GI: "'*AY* contains *B*' is the particular affirmative proposition with respect to *A*".

[17] §23 GI referred to here by Leibniz contains the rule of introducing a concept variable functioning as an existential quantifier: "For any definite letter there can be substituted an indefinite letter not yet used". (P, 57)

[18] Cf. C, 229—231, esp. the proof of 'Axiom 7'. A detailed discussion of this passage may be found in my ''Non est' non est 'est non'', *Studia Leibnitiana* **18**, 1986, 1—37.

[19] P, 65 omitts this word which clearly indicates the universal character of the indefinite concept *Y*; cf. the original: "Unde si sit *BY*, et terminus indefinitus quicunque [!] sit superfluous . . . tunc *B* est *Individuum*."

[20] Some alternative definitions are discussed in section 3.3 of my book *Das System der Leibnizschen Logik*, de Gruyter, Berlin, 1990.

[21] Here as well as in some further details I deviate from Parkinson's translation which has 'exists' for Leibniz's 'est'. The point is that Leibniz goes on to distinguish the "essential" from the "existential" meaning of '*A* est' as '*A* est Ens possibile' vs. '*A* est Ens existens'. Therefore only the latter may justifiably be translated as '*A* exists'!

[22] Cf., e.g., AV, 79: "The universal negative proposition 'No *B* is *C*' is reduced to this universal affirmative one: 'Every *B* is not-*C*'".
[23] In later works Leibniz preferred the following variant where 'c' is eliminated, according to **Th6**, in favour of '=': "Univ. aff. '*A* = *AB*', part. neg. '*A* not = *AB*', Univ. neg. '*A* = *A* non-*B*', part. neg. '*A* not = *A* non-*B*'" (C, 236), i.e.:

Schema 3

U.A.	$A = AB$	$A = A \sim B$	U.N.
P.A.	$\neg(A = A \sim B)$	$\neg(A = AB)$	P.N.

[24] In 'Zur Einbettung der Syllogistik in Leibnizens 'Allgemeinen Kalkül'', *Studia Leibnitiana Sonderheft* **15** 1988, 38—71, I have defined a schema to be homogeneous if the formalisations of the 4 categoricals "follow the same logical idea", i.e. if they obey to the laws of obversion and opposition.
[25] Of course, **Th14** $I(A) \rightarrow (P(A) \leftrightarrow A\,\mathrm{c}\,0)$ holds as well.

KEVIN MULLIGAN

COLOURS, CORNERS AND COMPLEXITY: MEINONG AND WITTGENSTEIN ON SOME INTERNAL RELATIONS*

1. INTRODUCTION

'There is a philosophical question as to what one *really sees*"[1]. Wittgenstein's remark raises all sorts of questions: Does one see tables and chairs, people jumping up and down, their jumps, their sadness? Does one see colours and forms, coloured forms, dynamic and static, that are above or to the left of other coloured forms? If the latter, are these things one sees private entities or public entities as are, presumably, tables and chairs? If both answers are legitimate (sometimes, or whenever we see?) what are the relations between the people we see and the coloured forms that we also see? In other words, is what is presented to me in my visual field private, public or partly private and partly public?

In what follows I studiously avoid all such questions in order to concentrate on a prior matter. What sort of relations obtain between colours and between orientation properties? On nearly all answers to the questions in the last paragraph such entities do occur at least sometimes in our visual fields. In particular I shall be looking at what the two Austrian philosophers, Meinong and Wittgenstein[2], have to say about *internal relations* between colours and between orientation properties. A third type of such entities, perceived spatial form, was the subject-matter of many Austrian discussions and also fascinated Wittgenstein. But I shall ignore it here[3] just as I shall bracket the question whether the redness of the object I see, or its being to the left are or are not phenomenal or mind-dependent properties.

The descriptive phenomenology of colours and other qualities I shall expound and explore in what follows is that set out by Meinong in his 1903 paper 'Remarks on the Colour Solid and the Law of Mixtures' and by Wittgenstein in lectures and writings in 1930, in particular in his *Philosophical Remarks*. Meinong's paper was published only six years after his pupil Höfler had described the colour solid in his *Psychologie* and is the most thorough account of the philosophical issues raised by this representation of the structure of colours.

Wolfgang Spohn et al. (eds.), Existence and Explanation, 77–101.

Although, as we shall see, there is considerable agreement at the descriptive level between Wittgenstein and his Austro-German predecessors, Wittgenstein's claims about his descriptions differ radically from the claims made by his predecessors such as Meinong about the same descriptions. Here I shall mention only two such differences. First, Wittgenstein's descriptions serve a number of purely practical purposes; those of Meinong and the other descriptive psychologists such as Husserl and Stumpf have a purely theoretical goal. Meinong, unlike Wittgenstein, took his account of the internal relations amongst colours to provide a scientific foundation for the empirical theories of colours developed by empirical psychologists. Wittgenstein, of course, had no such ambition although, like his Austrian predecessors, he did think that description was logically prior to empirical explanation and accounts of the causal genesis of phenomena (cf. PR § 1).

Secondly, Meinong and Wittgenstein had very different views about the status of the sentences that deal with internal relations amongst, for example, colours. Meinong introduces his distinction between apriori and empirical propositions with the contrast between, on the one hand, the three propositions

> Red is different from green
> Grey lies between black and white
> Orange lies between yellow and red

and

> The stone is falling

The first three, unlike the fourth, are, he says, necessarily true (UEW § 1). For Meinong descriptions of internal relations amongst colours are made true by the obtaining of necessary relations. The dimensions of colour space are 'natural' (BFM 522). For Wittgenstein, of course, the obtaining of internal relations and properties cannot be asserted (TLP 4. 122), a negative point to which Wittgenstein was to remain attached even after he had changed his mind about what was to count as internal. We cannot justify rules of grammar by sentences like 'There really are four primary colours' because such rules are arbitrary (Z § 331); a colour system does *not* reside in the nature of colours (Z § 357, LWL 8, 12). A full account of this difference between Meinong and Wittgenstein would of course involve examining the theories of internal relations and of necessary truths developed by the descriptive psychologists and the criticisms of positions like theirs by Schlick and Wittgenstein in their defences of the view that genuine propositions always describe contingent situations.[4] Such questions will also be bracketed in what follows.

In spite of important differences such as these it is a striking fact that the one recurring theme in Wittgenstein's philosophy for which little in Frege or Russell prepare the reader and which was to mean little to his successors is the nature of internal relations in general and of formal concepts in particular — just the theme that recurs in the writings of Wittgenstein's Austrian predecessors, Husserl and Meinong.

In § 2 I outline the now unfamiliar view that there is a purely formal notion of space of which the internal relations amongst colours provide one example. In § 3 some features of colour space according to Meinong and Wittgenstein are described. § 4 sets out the details of Meinong's views on internal relations amongst colours and amongst orientation properties. § 5 mentions three applications of Austrian views of internal relations.

2. COLOUR SPACE AND COLOUR GEOMETRY

The *Philosophical Remarks* refer to a remarkable number of different *spaces*: colour space (§ 1), auditive space (§ 42), tactile space (§ 214), pain space (§ 82), visual space (§ 206), kinaesthetic space (§ 73), a space to which both memory and reality belong, the spaces of movement (§ 140) and of orientation (§ 207) and the dark-light space (§ 45) — to mention only a few. The project of describing such manifolds goes back to the work of such heirs of Brentano as Stumpf, Husserl and Meinong who, taking up some remarks of Riemann, envisaged descriptions of the manifolds that constitute mental states and acts — kinaesthetic and perceptual states, judgement and supposition, etc., — of the manifolds of the (possible) objects of such acts as well as descriptions of the manifolds embracing both of these. Like Wittgenstein, the descriptive psychologists, up to and including the Gestalt psychologist-philosophers, envisaged descriptive investigations of the internal relations amongst colours as well as tones, harmonies and even electrical charges (PR §§ 81, 3–4).

Meinong in particular had described such claims as that 'colours constitute an at least three-dimensionally extended manifold (*Mannigfaltigkeit*)' as belonging to 'colour geometry', a discipline whose 'cognitive dignity' would, once developed, be quite comparable to that of geometry (BFM 408–9). What Wittgenstein in PR called the colour system and the grammar of colour is later called, en passant, colour geometry (BF I § 66, BF II § 86; cf. Z § 346, BF III § 3, § 86, § 154). But whereas the descriptive psychologists aimed to describe a variety of different spaces in terms of a theory of manifolds and variations along interdependent dimensions the concept that does all the work for Wittgenstein is that of a scale or yardstick (*Massstab*, PR § 82), for example a colour scale ('Yardstick and Propositional System', in PR); it is the applications of such a conventional system that underpin the — perhaps only metaphorical — extension of the concept of space to the domain of colours and elsewhere.

Colour geometry, like the grammar of colours, describes internal relations between colours. An internal relation, in contrast to an external one, is a relation such that if its terms exists then the relation obtains.[5] What, then, are the terms of internal relations?

One important negative feature of such platitudes as the example already mentioned, 'Orange lies between yellow and red', was noted by Meinong: such relations do not describe relations between things. But Meinong's positive account of the terms of such relations is unclear: he says that colour geometry deals with the nature (*Beschaffenheit*) not the existence of objects (BFM 499, UA 193, UEW 10) indeed the existence or non-existence of objects that have such properties is irrelevant when one is describing necessary or internal relations (Meinong (1882), 38; UEW § 1). But elsewhere Meinong claims that colour geometry is a branch of the theory of objects, of ideal incomplete objects (Meinong 1907, 121, 134). And then again, after correctly distinguishing between the colour green and green things, he tells us that the former is an ideal property (Meinong (1906), 26–7)[6], a view shared by Höfler (Höfler (1897), § 22).

When Wittgenstein talks of 'internal properties and relations' in the *Tractatus* he gives the example of two blue colours standing in the internal relation of brighter than/darker than. But he calls these two colours 'objects' and goes on to add, helpfully, that his 'vacillating use' of 'property' and 'relation' corresponds to his vacillating use of 'object' (4. 123). Although the view that there is a philosophically clear distinction between predicates and subjects, properties and their bearers, is heavily criticised in PR, grammar is, we are told at PR § 7, a "theory of logical types". And when Wittgenstein discusses colour incompatibility he does not hesitate to talk about incompatibility between different degrees of a property (PR § 80, cf. § 207). Much later, at the beginning of his *Remarks on Colours*, Wittgenstein writes:

A language-game: Report whether a certain body is lighter or darker than another. But now there's a related one: State the relationship between the lightnesses (*Helligkeiten*) of certain shades of colour (. . .) The form of the propositions in both language games is the same: "X is lighter than Y". But in the first it is an external relation and the proposition is temporal, in the second it is an internal relation and the proposition is timeless. (RC I § 1)

Meinong and Wittgenstein, then, seem to think that internal relations obtain amongst non-things. And for many ontologists this is also to claim that they obtain between entities that do not change. If we try to get a positive account of the terms of internal relations we run up against their vacillating use of 'property' and 'object'. In Meinong's case the vacillation is probably due to the fact that he never managed to provide a clear account of the relations between incomplete (variable) objects and properties. (The same is true of Husserl.) Nevertheless Meinong is quite clear that internal relations amongst colours are not relations amongst psychological entities, indeed he considered his account of colour geometry to be one of his main credentials as an opponent of Psychologism: 'colours are as little psychological as places or numbers' (BFM 498). Colours and their relations are merely the possible objects of psychological states.[7]

Are there any internal relations amongst things? If Kripke is right about identity, then numerical difference would be such a relation. Meinong suggests that numerical difference between colours is an internal relation: its terms are fundamenta, and if these are 'different then they *must* be different, where this word has the sense of 'logically necessary' '. And his formulation suggests that numerical difference is always a necessary relation, i.e. even if it obtains between two things (Meinong (1899), § 7). But this seems to be the only example of an internal relation between things that is entertained by Meinong.[8]

Do internal relations hold only between ideal entities — objects or properties? Are the only entities that do not change ideal entities? There *is* an ontological category looked on with favour by the early Meinong and Husserl that yields candidates for the role of terms of internal relations which are temporal and non-repeatable: the category of *moments*, or individual qualities, the redness of this cheek which is numerically distinct from although exactly similar to, the redness of some other cheek.[9] The category of moments is particularly attractive for the nominalist since it allows him to give an account of internal relations that dispenses with Platonic terms for these relations: each of the two rednesses is as temporal and unrepeatable as is the cheek in which it inheres, yet between these two rednesses there obtains the internal relation of being as dark/light as.

In spite of their fundamental lack of clarity as to the formal type to which the terms of internal relations belong, our two Austrian philosophers share the conviction that the internal relations amongst colours are best represented by the colour octahedron. 'An octahedron with the pure colours at the corner points e.g. provides a *rough* representation of colour-space (. . .) Using the octahedron as a representation gives us a *bird's eye view* of the grammatical rules.' (PR § 1) As Meinong's pupil Witasek puts it, this spatial representation of the 'system' of empirical colours provides us with the best 'bird's eye description' we have (Witasek (1907), 148).

Such a grammatical representation 'tells us what we can do (. . .) But grammar is not entirely a matter of arbitrary choice. It must enable us to express the multiplicity of facts, give us the same degree of freedom as the facts.' (LWL 8) It should be noted that although Meinong and Wittgenstein differ on the question whether the basis of the internal relations they are interested in is conventional, they both stress that the geometry of colour allows for, indeed requires what Meinong calls 'independent variability' within each of the dimensions in a given manifold with respect to the other dimensions (BFM 499).

What does such a bird's eye view, an *übersichtliche Darstellung*, represent? We have considered the question as to which *formal* categories the entities so represented belong (ideal objects or properties, non-ideal or temporal moments). But what *material* category do they belong to? Does 'colour' mean pigment, the colours we see and imagine, or possible colours no one has seen?

Both Meinong and Wittgenstein are emphatic about the distinction between 'physical' colours, or pigment, on the one hand, and 'psychological' and ideal colours on the other hand and the corresponding distinction between two types of colour mixtures (BFM §§ 11—18; PR § 218).

One aspect of the distinction between empirical and apriori truths about colours that exercised Meinong's attention is the question whether the colours we sense and imagine exhaust the domain of colours. He first considers the parallel question about tones that is also discussed by Wittgenstein: does the musical scale carry within it an infinite possibility of being continued?

The finitude of the musical scale can only derive from its internal properties. For instance, from our being able to tell from a note *itself* that it is the final one, and so that this last note, or the last notes, exhibit inner properties which the notes in between do not have. (PR § 223)

Meinong too thinks that quality series are 'in principle limited': the last member of the scale is 'something that is psychologically given'. But he immediately introduces a distinction that can make no sense for someone who, like Wittgenstein, believes that the system of colours or tones does not reside in the colours and tones themselves: the real, psychologically given colours are in principle limited, not so all possible colours (BFM 505).

Our knowledge of colour space is as apriori in nature as our knowledge of real space: it is colour geometry. Our knowledge of the colour solid is empirical in nature and to this extent colour psychology: but apriori elaboration of what has been obtained empirically is as little excluded here as elsewhere in the empirical sciences. (BFM 534, cf. 507f.)

(Meinong mentions theoretical mechanics as an analogous case). It is because of this distinction between actual, psychologically real and merely possible colours that Meinong is prepared to entertain the possibility that colour space may exhibit more than three dimensions.

The main types of internal relation between colours distinguished by Meinong and Wittgenstein are: the relations of order and distance and the special cases of kinship or similarity and contrast. (We shall look at another relation, incompatibility, in § 4 below).

Consider the position of the complementary or contrast-colours on the colour solid:[10]

We have apriori insight into contrast in the sense of opposition (*Gegensätzlichkeit*) in the case of the colour tones. That a certain red it opposed to a certain green as the colour at the greatest distance from it, and likewise for a certain yellow and a certain blue, such that all other colour tones exhibit a lesser difference of distance — all this I can see within certain modes limits of reliability as following from the nature of the different colour tones and to this extent apriori. (BFM 501)

'One could say that the colours have an elementary affinity with one another'. (PR § 76) Later, in the *Remarks on Colour*, Wittgenstein says of the relations of kinship and contrast among colours: "that is logic" (RC II § 46).[11]

3. THE COLOUR SQUARE VS. THE COLOUR CIRCLE THE *JUSTE MILIEU*

Meinong and Wittgenstein both thought that the colour octahedron, and in particular the square with the four corners corresponding to red, yellow, green and blue, brings out particularly well some of the internal relations of order amongst colour.

There is no lack of insights concerning sameness and difference of direction on the colour solid that can without any hesitation be called apriori: that the connecting line from red and orange leads to yellow, that the line from red and violet leads to blue, but that these two lines do not by any means form a straight line (*gestreckter Winkel*, an attenuated angle; BFM 501).

Similarly, Moore reports that in 1930 Wittgenstein was claiming that the arrangement of the four primary saturated colours in a square is superior to their arrangement in a circle since it brings out more clearly the 'logical' distinction between those saturated colours that are and those that are not primary (Moore (1959), 317—8). This point is developed in a a subtle way in the *Philosophical Remarks* and in other records of Wittgenstein's lectures at the beginning of the thirties.

The difference between the square and the circle is the difference between the central cross-section of one of the different versions of the octahedron developed by Höfler and Ebbinghaus, on the one hand, and the equator of Runge's globe (*Farbenkugel*), on the other hand. As Höfler had pointed out, the disadvantage of the latter is that it represents the relative distances between colours in an arbitrary and very probably incorrect fashion. Whereas the square arrangement brings out "certain other inner relations (*innere Beziehungen*)", in particular the relations of opposition (Höfler (1897) § 24).[12]

The square arrangement rather than the circular arrangement brings out the difference between two sorts of betweenness. The first type of betweenness is illustrated by

Any purple is between pure red and pure blue
Any orange is between pure yellow and pure red

the second type by

Any pure red is between orange and purple.

That there is a difference here, Wittgenstein says, can be expressed by saying that 'whereas an orange can properly be called a 'mixture' of yellow and red, red cannot properly be called a 'mixture' of orange and purple.' (Moore 1959, 317) The square arrangement, we may suppose, is superior to the circular arrangement because on the square, as opposed to the circle, there are privileged points that correspond to the privileged position of the primary saturated colours. The square arrangement allows us to see what we might call the juste milieu.

As Wittgenstein points out we might choose mixed colours at random and stipulate that they include an angle of 90°. But this would be completely arbitrary, 'here the *metaphor* of an angle collapses' (PR § 221).

We say a colour can't be between green-yellow and blue-red in the same sense as between red and yellow, but we can only say this because in the latter case we can distinguish the angle of 90°; because we can *see* yellow and red *as* (my emphasis) *points*. (PR § 221)

. . . we cannot say that red has an orange tinge in the same sense as orange has a reddish tinge. that is to say, it seems to be clear that the phrases "*x* is composed of (is a mixture of) *y* and *z*" and "*x* is the common component of *y* and *z*" are not interchangeable here: Were they so, the relation *between* would be all we needed for a representation. (PR § 222; cf. Hofmann (1913), 48)

4. VISUAL SPACE AND COLOUR SPACE

The central difference between the two Austrian accounts of colour space is that Meinong, unlike Wittgenstein, takes seriously the notion of space that is appealed to in talking of tone space, colour space, etc. This leads him to give a much more thorough – but also more ambitious – analysis of such concepts as *simple* and *complex, mixture, incompatibility* than Wittgenstein. Where Wittgenstein arrives at conclusions that resemble those of Meinong he, unlike Meinong, thinks that he is pointing out features of the (at least partly) conventional application of coordinate systems.

If we take seriously the claim that colours are in colour space and the claim that a colour line or a colour surface do not do justice to the colours then we may say with Meinong that the colour octahedron is a colour body or solid (*Farbenkörper*, BFM § 2) in colour space, an at least three dimensional entity whose job is to "include empirical colour data before all theory" (BFM 534). 'The colour body is in colour space and participates in its properties.' (BFM 534) Now Meinong's analysis of colour space involves two features that are controversial. First, it is designed to fit the Hering four colour theory. Meinong recognises that this might be thought to conflict with the requirement that description ('phenomenology') be theory free. Wittgenstein, too, seems to assume that the four-colour account of Hering is superior to the different three-colour accounts. Secondly, it involves an unusual account of the nature of the dimensions of colour space. This revision of the traditional view of the dimensions in which colours can vary was held to be unpalatable even by Meinong's own pupils, Höfler and Witasek. Nevertheless, as we shall see, in spite of the controversial details of its application to colour space, Meinong's general analysis really does provide some of the basic features of any approach that takes 'spaces' (ontologically) seriously.

I shall describe first, the central claim of Meinong's analysis; second, the main problem on which the analysis is designed to shed some light; third, the application of the main claim to the relatively uncontroversial case of the space of visual orientation; fourth, Meinong's unusual view of just what the dimensions of colour space are and his arguments for this view; finally, the application of the main claim to colour space.

Meinong's central *claim* is a perfectly general point concerning manifolds and objects, simple and complex. The members of an n-dimensional manifold cannot possibly be simple.

> If an A and a B can vary in the same two dimensions then this involves the possibility that A and B will be found to be the same in one respect and not in another: two simple objects cannot possibly be simultaneously different from one another and the same. There are as many components as there are dimensions . . . . (BFM 515)

If colours are in a three dimensional colour space then we may call the three components of every colour 'colour elements' (*Farbenelemente*). In precisely the same way, Meinong points out, we can introduce the concept of 'spatial elements' (*Raumelemente*) in our analysis of visual space with its three dimensions, above-below, left-right and in front-behind (cf. PR § 206).

The *problem* on which this claim is supposed to shed some light is the problem of incompatibility: just as an account of colour elements is going to help us understand the impossibility of sensing red and blue at the same place and at the same time so too the account of spatial elements is going to help us understand the parallel impossibility of seeing one and the same object at the same moment as directly in front and to the right.

Incompatibility amongst colours is the starting point for the discussions of colour by both Meinong and Wittgenstein,[13] it is a relation both take to be an example of an internal relation:

> One's first thought is that it is incompatible for two colours to be in *one* place at the same time. The next is that two colours in *one* place simply combine to make another. But third comes the objection: how about the complementary colours? What do red and green make? Black perhaps? But do I then see green in the black colour? – But even apart from that: how about the mixed colours, e.g. mixtures of red and blue? (PR § 76)

Meinong's thoughts, he tells us, had followed a similar sequence. Before 1903 he had defended the view that "it is as impossible to see at the same position and at the same time red and blue, or even to imagine them, as is a round square" and "that no analysis is capable of finding pure red and pure blue in violet" for "in violet one does not have both red and blue but neither red nor blue, rather a third thing lying between red and blue". In his 1903 paper however he presents an analysis of colours that, he claims, brings out the hidden truth behind the claim that violet is a mixture containing red and blue. This analysis is presented not only for the case of colour properties but also for the case of orientation properties.

4.1 *Orientation Space: Corners*

Let us therefore see how Meinong's analysis applies to orientation properties.

Suppose I stand in a square room with my back to one of the walls. Can one say

> that the the corner of the room in front of me and to my right combines the place properties exhibited by a certain position on the wall in front of me and a position on the wall to my right? If I put the question like this the incompatibility of the two properties becomes immediately evident.... (BFM 517)

Meinong baptises the different spatial elements here 'depth data', 'breadth data' (not to be confused with length) and 'height data'. If we assume, in order to simplify, that all the positions in our example involve the same height datum then Meinong's description of our example runs as follows: *in front of me* means a certain depth, *t*, together with a certain breadth value, *b; to my right* means a certain breadth, *b′* and depth, *t′*. The right-hand corner in front of me is then characterised by *t* and *b′*, but not by either *b* and *t*, or *b′* and *t′*.

As Meinong notes, someone who finds it strange to characterise a position on the wall which is directly in front of the subject as having a breadth value, someone who finds it strange to say that what is neither to the left nor to the right has a position on the left-right dimension, is confusing the coordinate value zero with the absence of a certain property. Wittgenstein makes the same point in a passage on money-space and pain-space: " 'I haven't got stomach-ache' may be compared to the proposition 'These apples cost nothing'. The point is that they don't cost any money, not that they don't cost any snow or any trouble. The zero is the zero point of *one* scale." (PR § 82)

What is incompatible are the two *complexes*, (*b*, *t*) and (*b′*, *t′*), although their components, *t* and *b′* are compatible (BFM 518). Incompatibility as a relation between complexes does not preclude compatibility between the *components* of these complexes.

4.2 *The Dimensions of Colour Space*

Before we look at how the analysis carries over to colour space Meinong's unusual account of the *dimensions* of colour space must be introduced. Nothing would seem to be more natural than to take the dimensions of colour space to be the three dimensions in which colours can vary: hue, saturation and brightness. This was certainly the view of Höfler, Witasek and Wittgenstein. But, objects Meinong, hue cannot be a dimension since 'The changes in colour hue take place in two dimensions' and 'there is no room in *one* dimension for a closed line'. Meinong never mentions the possibility that the privileged points which are the primary colours could be represented without employing cartesian coordinates (for example with polar coordinates.) Saturation cannot be a dimension either since, if grey really belongs to the centre of the colour solid, then saturation too will vary in two dimensions. Brightness is, however, a dimension, but not merely a dimension of the colour solid but of colour space (BFM 508). Hence the bright-dark dimension does not coincide with the white-black line, although the latter is in this dimension. The white-black line is the 'main representative of (proxy for)' the brightness dimension.

Meinong then argues from the point that Wittgenstein was also to make

> ... no-one could see both orange and violet in pure red in the way in which it has so often been thought that red and blue can both be found in violet. (BFM 512)

to the claim that red and green, on the one hand, and yellow and blue on the other hand, are the main representatives of the two other dimensions of colour space. Normally, Meinong suggests, the names 'red', 'green', 'yellow' and 'blue' designate such main representatives. Only in one type of case do they seem to be used to designate directly a dimension: when we refuse to call the grey point red although other points at the same distance as it is from the red corner are called red.

The *application* of the general analysis to colour space in order to give an account of *colour incompatibility* now turns out to be very simple. What we know as pure red and pure yellow are just as incompatible as the complexes *b, t* and *b′*, *t′* in the above example. But this red and this yellow cannot be simple because of their positions in a three dimensional continuum

> ... and if, in the interests of greater simplicity, we disregard the colour elements required by the light-dark dimension there remains in each of these colours a red-green element, *r* or *r′*, and a yellow-blue element, *b* or *b′* (...) Of course *r* and *b′* are compatible with one another. (BFM 518)

What, then, on this analysis is a so called mixed colour? A psychological colour mixture is *not* the coming together of pure red and pure yellow; orange, for example, involves only the two elements *r* and *b′*. 'The ordinary point of view, which overlooks the elements *b* and *r′* in pure red and yellow will not hesitate to take the elements *r* and *b′* to be pure red and pure yellow.' (BFM 518) Instead of talking of (psychological) colour mixtures it is always better to use the relation already mentioned of betweenness: "in violet one does not have both red and blue, but neither red nor blue, rather a third thing lying between red and blue" (BFM 514); "a mixed colour, or better, a colour intermediate between blue and red" (PR § 80).

Meinong's account of what the dimensions of colour space are denies to saturation the rôle of a dimension but it does allow him to analyse saturation. The 'neutrality values' of the red-green dimension and the blue-yellow dimension together yield grey, and when combined with the neutrality value of the brightness dimension, neutral grey. In the case of a pure colour such as red, which is characterised by an extreme value of the red-green dimension and the neutral value of the blue-yellow dimension, we have a tendency to refuse to recognize the presence of the neutral element in spite of its 'distinguishing function'. Extreme components seem to have a greater significance than neutral or nearly neutral values; they seem to be more 'intellectually accessible'. This, Meinong suggests, is what the concept of saturation describes (BFM 519).

On Meinong's account we can see why, as Wittgenstein was to put it, "a colour intermediate between blue and red is such in virtue of an internal relation to the structures of blue and red" and why "this internal relation is *elementary*. That is, it doesn't consist in the proposition '*a* is blue-red' representing a logical product of '*a* is blue' and '*a* is red' " (PR § 80). Every colour in colour space is characterised by a structure consisting of three values of the three dimensions of colour space — whichever view of these dimensions is adopted. Such a structure is a complex consisting of three such values or elements and between all such elements there obtain internal relations of three-way reciprocal generic dependence: every colour element requires two other elements from the other two dimensions. Every colour is specifically dependent on its constitutive elements, as a complex of such elements is specifically dependent on the colour they constitute. Between elements from the same dimension there are internal relations of order and distance and similarity. Incompatibility is another formal relation but a relation that holds between complexes. In other words, there are more ways for elements to hang together than are dreamt of in the *Tractatus*.[14]

In the *Tractatus* Wittgenstein had characterised the dependence of an internal relation on its terms in terms of what is and is not thinkable: the internal relation darker/lighter than between two blue colours is such that it is unthinkable that these two objects do not stand in this relation (4. 123). But as Meinong (and Husserl) point out, that something cannot be thought of without something else, must be distinguished from something's not being able to exist without something else (BFM 505). And this distinction is important for Meinong because of his distinction between the colours we actually see and possible colours (see § 2 above).

4.3 *The Nature of Elements*

Meinong points out that in the absence of a precise account of the nature of those constituents which are colour elements and of the way they combine there is an unfortunate tendency to rely unconsciously on the analogy of material parts (BFM 520). But one interesting clarification of the nature of Meinong's elements does follow from what he says. The elements of the space of orientation must, it seems, be properties, such as sinistrality, centrality, or representatives of properties. The two cases of the spaces of colour and of visual forms are much more difficult, as we have seen. But a distinction often used by Meinong and other pupils of Brentano is useful, and one half of this distinction is actually criticized by Wittgenstein. The distinction is that between the ideal objects *Red* and *The Triangle* on the one hand, and their instantiations on the other hand. The instantiations of Red, etc. are the dependent particulars already alluded to as the only temporal entities between which internal relations can obtain, often called *moments* by the Brentanians ('qualities' and 'tropes' by others). Of such moments the Brentanians often liked to say that they were 'abstract parts' of objects. (The term 'qualities' has often been reserved for what is, like Aristotelian accidents, 'in' an object or a subject without being a material part of the object). Similarly, the instantiations of geometrical forms were thought of as dependent particulars, two and three dimensional *Ausdehnungen* or extents. *Instantiation* as a relation between ideal objects and dependent particulars is to be sharply distinguished from *exemplification*, the relation between an ideal property and particulars and the case where an ideal form is instantiated by a concrete, independent form.[15]

Wittgenstein criticizes the idea that a 'red circle *consists* of redness and circularity or is a complex of these components as a misuse of these words and misleading. (Frege knew this and told me so)', in the appendix to PR on 'Complex and Fact' and later.

Nevertheless moments or 'abstract parts' have at least three important roles in the accounts of perception of Husserl and the early Meinong, rôles that perhaps only they can perform. First, together with the independent particulars which are instantiations of ideal forms they are, on these accounts, what we see. Thus moments allow us to defend the thesis that what we see is always particulars, independent and dependent, relational and monadic. And, secondly, they are what we often designate with the help of such singular terms as 'the redness of that cheek'. Finally, the internal relation of interdependence between those perceptual objects which are colour moments and their associated extents allows us to distinguish between the *complexity* of a colour moment (its constituent colour elements) and its *uniformity* – the fact that a particular extent is uniformly red.[16]

5 DEVELOPMENTS AND APPLICATIONS

The descriptions given by Meinong and Wittgenstein of the internal relations which, together with external relations such as contiguity and succession, structure what we see represent a point of view which has become relatively uncommon. At the opposite theoretical extreme is the point of view developed by Russell's pupil Jean Nicod in his wonderful *La Géométrie dans le monde sensible* (1924) which provides a taxonomy of sensory relations all of which are understood as external relations. It still remains to be seen which of these two accounts is the most adequate. Here I shall limit myself to pointing out three developments and applications of the Austrian view of colours and corners which any such evaluation would have to take into account.

5.1 *Tractarian Complexity*

As Meinong's distinction above between simple and complex objects indicates, the discussions of colour by Meinong and Wittgenstein may well throw light on some of the concepts employed in the *Tractatus*: independence and dependence; simple and complex; and formal relations, since all internal relations are formal.[17] Wittgenstein's discussion of colours in PR is indeed presented as pointing towards a fundamental modification of some Tractarian doctrines:

> In my old conception of an elementary proposition there was no determination of the value of a co-ordinate; although my remark that a coloured body is in a colour-space, etc., could have led me straight to this. (PR § 83)

On the Husserlian and Meinongian accounts of the manifold of propositional form, of the manifold of the form of states of affairs and of the manifold that embraces these the *Tractatus* represents a deviant version of manifold theory. Deviant, because Wittgenstein insists only on independent variability within the different dimensions distinguished and provides almost no account of the interdependence between dimensions, a gap that is partially remedied in the *Philosophical Remarks.*

5.2 *The Variety of Colours*

The colour octaedron tells us a great deal about the phenomenology of colours, but by no means everything. It tells us nothing about the rôle of illumination. and hence nothing about the colours we encounter most frequently. After PR Wittgenstein seems to have discovered the variety of colours. Within descriptive psychology the different types of colour not represented by the colour solid had been described and analysed by David Katz in his classical 1911 monograph *Die Erscheinungsweisen der Farben.* As Katz puts it, colour space has traditionally been conceived as dealing with *Flächenfarben* or film colours; but none of the stereometric constructions of colour solids can be considered to be a conformal mapping (*konforme Abbildung*) of all the *Oberflächenfarben* or surface colours (Katz (1911), 35). It is via *Oberflächenfarben* that we get information about the external world (Katz (1911), 266). A full description of colours must start from this distinction and describe such phenomena as space colours (e.g. the colour of fog), the colour of stuffs as well as the phenomena of transparency, lustre and reflections.

Wittgenstein had already accepted in PR two claims (characteristic of the descriptive psychologists) without which it is imposible to describe these varieties of colour. First, the claim that what is seen is coloured and spatial (PR § 205ff.) It is probably to this claim that he is alluding when he says that colours may have, in addition to saturation, hue and intensity of light a fourth dimension[18] (PR § 66). Second, the further specification of this claim: that we see depth (PR § 208). But the descriptions of the variety of colour in *Remarks on Colour*, which build on these two points, differ in one crucial respect from those of Katz. Wittgenstein's moral is, from the theoretical point of view, negative: not all colour concepts are of the same kind, the concept of sameness of colour is indeterminate, etc. Katz, on the other hand, is concerned to set out the relations between the different systems of colour.

5.3 *Internal Relations and the Analysis of Perception*

Finally, perhaps the most important job performed by the view documented above is within a certain theory of perception, a theory developed by the heirs of Brentano, above all by Husserl in *Ding und Raum* — which describes the internal relations within kinesthetic space and orientation space — and by Paul Linke. It contains two major claims, each of which is anti-Kantian. It therefore differs considerably from analyses of perception that stem from Kant such as those of Helmholtz and Poincaré.

The first component in this Austrian view of perception is the thesis that to see is not to judge and hence involoves no exercise of concepts. This view was defended by Husserl and by Meinong's pupil Benussi and has recently been forcefully defended by Dretske.[19] Meinong himself never relinquished the thesis of his teacher Brentano for whom to see is to judge, but since Meinong thought that perceptual judgements worked with intuitive presentations rather than non-intuitive presentations, concepts, his view is after all closer to Husserl's than that of Kant. Since first person perceptual reports are founded on perceptual judgements (when they are not *Ausserungen* of perceptions) this view claims that such judgements differ fundamentally from their perceptual bases. Of course, this claim is quite compatible with its being the case that what is seen depends functionally or in other ways on the concepts a subject has mastered (on the bases of his dispositions to employ different concepts).

The second component of this Austrian view is the already mentioned claim that a spatial element, in the simplest case a visual point or extension, but normally a three dimensional body, always belongs to what is seen.

This Austrian view must of course, like any analysis of perception, account for what might be called our ability to identify and discriminate items within the scenes we are visually presented with. In order to do this it cannot appeal to the employment of concepts, as can the friend of the Kantian approach. The view that the colours, forms and orientation properties we are presented with are highly structured in virtue of the internal relations holding between the different features within a visual field does just this job. Just *how* it does this job will depend on the answers given to the questions with which we started. Thus the coloured forms we see may be held to be private phenomenal entities standing in one or more relations to public things: belonging to, caused by, manifestation of, etc. Or they may be held to be public entities that stand in one or another part-whole relation (three-dimensional part, abstract part) to other public things. Or they may be held to be public entities that are identical, at a time, with people and tables.[20] Let us see how the Austrian view works on the assumption that the latter view is the correct one (translations into the language of the other views are easily effected).

On the Austrian view the perceptual subject tracks, stands in perfectly extensional relations to, coloured forms that are static and dynamic, relational and non-relational. What perceptually individuates all such items are the external and internal relations they stand in, not the concepts under which they fall. Now it is a striking characteristic of the perceptual reports that are based on — but not, on this view, identical with — perception that their primary use is to describe external relations amongst things:

(1) Sam sees Maria

(2) Sam sees Maria hit Hans

(3) Sam sees Maria hitting Hans

(4) Sam sees Maria fall over the stone

What makes such descriptions of external relations true is Sam's standing in perceptual relations to coloured forms and the relations, internal and external, amongst these and their parts. Perceptual reports *may* of course describe internal relations

(5) Sam sees this ice patch as darker than that patch

may be made true by Sam noting an internal relation between two colours rather than an external relation between the two patches.

Wittgenstein describes the difference between the two interpretations of (5) in the *Last Writings on the Philosophy of Psychology*:

> "I see that both of them are similar" can be used temporally and atemporally, depending on how "both of them" are defined . . . "I see" is always temporal, but "both of them are similar" can be atemporal. (LWPP § 152)

'But do I therefore see something different in each case?' Wittgenstein asks. On the Austrian view, the answer to this is: no, although there are differences in attention between the two cases. Not only the truth maker of (5) but also the truth makers of (1)–(4) involve perceptual relations between Sam, on the one hand, and internal relations between coloured forms on the other hand. Even in the case of (1) Sam is in perceptual contact with a complex whole and with internal relations between its parts. He sees Maria's arm as belonging to the shape that is at that moment Maria.

Wittgenstein pursues the difference between reports that are and are not explicitly about internal relations as follows:

> "There is similarity between these two cases" . . . could mean: there is a similarity between *this* type of face and *this* type of face . . . It may be the faces of *these people* that interest me, or these facial *forms*, wherever I come across them.
>
> The difference I have in mind is of course that between the sense: These two lines have a similar form — and the sense: the circle, ellipse, parabola and hyperbola resemble one another. (§ 155)

The difference is that between 'internal and external similarity'. In the former case the relation is between geometrical types (§ 156, § 158). (As we saw above, in PR grammar deals with types.)

Internal relations strike the subject, come to his attention, in the phenomenon of sudden seeing as ('Aha-experiences').[21] For in such unusual experiences the subject is struck by internal relations that are normally not thematic, not the object of his attention: he notes one coloured form as fitting or going with a second coloured form. But for Husserl and Meinong, Benussi, Linke and other Gestalt psychologists — but probably not for Wittgenstein — such surprises merely throw into relief the internal relations which always structure what we see. That the parts of the coloured array that is Maria belong together is a presupposition of ordinary 'continuous seeing'. This point is brought out by a a type of perceptual report introduced by the descriptive psychologists:

(6) Sam sees A as a part of B/as fitting or going with B/as brighter than B. . .

where these relations are internal relations.[22] The truth makers of such reports, which employ only colour and form expressions, make true both reports of thing-perception (cf. (1)), naked infinitives describing events and processes ((2), (4)), reports of perception of states ('Sam sees this animal as a duck' — not to be confused with the absurd 'Sam sees this duck as a duck' — and 'Sam sees Maria's sadness') as well as descriptions of the experience of aspect change.[23] Just how awareness of internal relations — which by itself tells us nothing about the world as long as the terms of the relation are taken to be ideal entities —

combines with sensory input in perception is the problem addressed by Paul Linke (1918) in one of the last developments of the descriptive psychology of perception.

One of the many possible objections to the claim that perceptual reports typically describe external relations whereas perception always involves awareness of internal relations is that perceptual reports, indeed all descriptions, also involve awareness of or mastery of internal relations, albeit conceptual ones. By way of reply it should be noted, first, that the claim that there are internal relations between concepts or ways of using words is a very different claim from the claim that colours, etc. stand in internal relations. The classical arguments against conceptual necessity do not apply obviously to the claim that there are internal relations amongst colours or tones. That our conceptual discriminations are conventional would not entail that the same is true of our perceptual discriminations unless it is shown that colours are linguistic entities. And with respect to Wittgenstein's apparent adoption of this view (see above), a view quite clearly not shared either by his Austrian predecessors, nor by many of the more interesting recent writers on perception (Harrison, Hardin, Westphal) Moore's astonishment is surely justified:

> According to what he said elsewhere, he could only have been talking sense if he was talking, not about the colours, but about certain words used to express them; and accordingly he did actually go on to say that "red is primary" is only a proposition about the use of the English word 'red', which, as I said . . . he cannot seriously have meant. (Moore (1959), 318)

Secondly, as the friends of semantic spaces are the first to point out, the number of dimensions of semantic spaces exceeds by far the number of dimensions in the different sensory spaces. Thirdly, it seems much more likely that a large part of our ability to recognize sensory facts is innate than does a similar claim with respect to our mastery of different semantic fields.

Although the Austrian analysis of perception goes well beyond what Wittgenstein says about perception the following formulation of Wittgenstein's neatly captures part of the strong Austrian view:

> I should like to say, if there were only an external connection no connection could be described at all, since we only describe the external connection by means of the internal one. If this is lacking we loose the footing we need for describing anything at all — just as we can't shift anything with our hands unless our feet are planted firmly. (PR § 26)

Perhaps all that Wittgenstein had in mind here is that a description and the state of affairs (external relation) it describes must stand in an internal relation to one another. On the Austrian account it is important to add that when we perceptually verify this description we see objects standing in both internal and external relations. We see external relations with the help of internal relations.

Department of Philosophy,
Geneva, Switzerland.

NOTES

* Thanks for many comments to Roberto Casati, from whom little about colours is hidden.

[1] Malcom's summary of Wittgenstein's lectures in 1946, Malcolm (1967), 49.

[2] No one has done more than Karel Lambert to bring out the coherence and interest of one of Meinong's best known theses. What follows will, I hope, help to show that there are more things in Meinong's philosophy than his account of definite descriptions and their objects.

[3] See Mulligan (1988, 1991); Smith (ed.) (1988).

[4] A proper appreciation of the difficulties here presupposes a grasp of the theory of materially necessary truths developed by Husserl in the third of his *Logical Investigations* and of the differences between this theory and the only superficially similar Kantian account of synthetic a priori judgements.

[5] This well known characterization of internal relations is to be found at Höfler (1897, § 22). Höfler refers back to Höfler/Meinong (1890) § 25 and § 55, which in fact merely anticipate the later formulations. The terminology of 'innere' and 'äussere Beziehungen' is Höfler's rather than Meinong's. (See also Höfler (1930) § 22, Höfler (1922) § 25.) Meinong marks the difference between the terms of an external relation and those of an internal relation by calling the former 'members' and the latter 'fundamenta' (Meinong (1899) § 7; cf. also Höfler/Meinong (1890) § 25).

[6] Meinong and his school became aware of the philosophical importance of the theory of relations around the same time as Russell, even though they never had a logic of relations. Thus Meinong's collaborator, the mathematician Zindler, had stressed the philosophical importance of the fact that relations can have more than two terms (on this, see Meinong 1899, 196). Höfler (1896) deals with complex relations.

[7] How Meinong thought this point should be reconciled with his late theory of perception, according to which colours cannot actually be seen, since they are not mind-independent, is a story that cannot be told here. Cf. UEW § 8.

[8] Wittgenstein at one point entertains the hypothesis that "the things themselves" include the four primary colours (PR § 147). This remark is particularly interesting for those who think that Tractarian objects correspond to what have traditionally been called properties, whether ideal or non-repeatable and for those who think that they correspond to ideal objects. On whether the colours discussed by Wittgenstein in

Remarks on Colour are properties or 'logical subjects', see Chisholm (1990) and Westphal (1987).

[9] On this category, see Smith (ed.) (1982).

[10] "Farbenkörper". Cf. "Beweiskörper, Bedeutungskörper".

[11] Cf. Husserl (1964) § 16.

[12] The primitive octahedron first developed by Höfler in 1897 does not bring out the way in which yellow is lighter than blue. But this is brought out in variations on the original octahedron by Höfler and Ebbinghaus, to which Meinong refers. In these irregular octahedra the distance from white to blue is about four times as great as that from white to yellow (Höfler (1911); RC I § 2, III § 132). As Höfler points out, the view that it is an inner feature of blue that it is dark, of yellow that it is light, etc. goes back to Schopenhauer.

[13] On the plausibility of colour incompatibility see Hardin (1988), 121f.

[14] In his discussion of perceptual spaces Hardin writes: "So the relations that the colors bear to one another are neither more nor less puzzling than the relations borne to each other by a set of vectors extending from a common origin in a space with four preferred directions" (Hardin (1988), 116). On the ontology of vectors, see Höfler (1900), Johansson (1989), 103f., 165f.

[15] Instantiation and exemplification must also be distinguished from the determinable-determinate and genus-species relations, internal relations the Brentanians liked to subsume under the rubric 'logical parthood'.

[16] Cf. Husserl's LI III § 8, Westphal (1977).

[17] On formal concepts in the *Tractatus* and the *Logical Investigations*, see Mulligan (1990a).

[18] Cf. Z § 269; Höfler (1930) § 24 defends this view against Helmholtz.

[19] Husserl, RL VI §§ 1—3; on Benussi see Stucchi (1991); Dretske (1969).

[20] This seems to be the correct view. It may even have been Aristotle's view: "those sense-objects are called incidental that are like the white thing's being the son of Daries. For this we perceive incidentally, for it is incidental to the white thing we perceive (*de Anima*, II 6). On this view, see Mulligan (1991).

[21] Cf. Chisholm (1989), Mulligan (1988).

[22] All the differences already mentioned about the nature, formal and material, of the terms of such relations and their scope crop up in the writings of the Gestalt philosopher-psychologists. Is a Gestalt such as a melody an ideal property (Ehrenfels, Meinong, Benussi), a dependent particular (Husserl, Stout), an individual (Köhler)? Are the parts of a Gestalt independent of it (Meinong, Stumpf) or inseparable from it (Köhler), that is, are perceptual relations exclusively internal? Discussions of the latter question suffered from the fact that all the participants, except Husserl, were critical realists, holding that Gestalten and their parts were first and foremost phenomenal entities. Once this phenomenalist assumption is discarded the way is open to claim (a) that the theory of relations of Meinong and Stumpf was a good one — relations ultimately presuppose independent terms — but these independent terms are not phenomenal items, for there are none such; and (b) that there are internal relations amongst the parts of the coloured arrays we see, but that these arrays and their parts are not phenomenal items or ideal entities but rather moments and things belonging to the same category as 'this square'.

[23] 'Seeing as', whether used to describe sudden aspect switches or not, takes as a complement a small-clause and resembles for this reason naked infinitives, which are used to describe events and processes rather than states.

REFERENCES

The abbreviations employed in the text and notes are given below.

Benussi, V.: 1914, 'Gesetze der inadäquaten Gestaltauffassung', *Archiv fur die gesamte Psychologie*, 396–419.

Chisholm, R.: 1989, 'Act, Content and the Duck-Rabbit', Proceedings of the 1988 Kirchberg Wittgenstein Conference.

Chisholm, R.: 1990, 'What are Wittgenstein's Remarks on Color about?', *Proceedings of the 1989 Kirchberg Wittgenstein Conference.*

Dretske, F.: 1969, *Seeing and Knowing*, Routledge, London.

Hardin, C. L.: 1988, *Color for Philosophers. Unweaving the Rainbow* Hackett, Indianapolis.

Höfler, A.: 1896, 'Zur Analyse der Vorstellungen von Abstand und Richtung', *Zeitschrift für Psychologie,* **X**.

Höfler, A.: 1897, *Psychologie*, Tempsky, Vienna and Prague.

Höfler, A.: 1900, *Studien zur gegenwärtigen Philosophie der Mechanik*, Pfeffer, Leipzig.

Höfler, A.: 1911, 'Zwei Modelle schematischer Farbenkörper und die vermutliche Gestalt des psychologischen Farbenkörpers', *Zeitschrift für Psychologie,* **58**.

Höfler, A.: 1930, *Psychologie*, 2nd ed., 1st vol. (only volume published) Tempsky, Vienna.

Höfler, A. (with A. Meinong): 1890, *Logik*, Tempsky, Vienna, second edition 1922.

Hofmann, H.: 1913, 'Untersuchungen über den Empfindungsbegriff', *Archiv für Psychologie*, **26**.

Husserl, E.: 1900/01, *Logische Untersuchungen*, 1st ed., Niemeyer, 2nd ed. 1913/21, (Husserliana XVIII, XIX/1, XIX/2); trans. by J. Findlay as *Logical Investigations*, London: RKP (LI).

Husserl, E.: 1907/1973, *Ding und Raum. Vorlesungen 1907*, Nijhoff, Husserliana XVI.

Husserl, E.: 1964, *Erfahrung und Urteil*, Claassen, Hamburg, (*Experience and Judgement*, Routledge and Kegan Paul, London, 1973).

Katz, D.: 1911, *Die Erscheinungsweisen der Farben*, second ed. 1930, *Der Aufbau der Farbwelt*, Leipzig; partial English trans. *The World of Colour*, Routledge and Kegan Paul, London, 1935.

Linke, P.: 1929, *Grundfragen der Wahrnehmungslehre*, Reinhardt, Munich, 1st ed. 1918.

Malcolm, N.: 1967, *Ludwig Wittgenstein. A Memoir*, Oxford

Meinong, A.: 1882/1971 'Hume-Studien II. Zur Relationstheorie', Gesamtausgabe Vol. II, Graz.

Meinong, A.: 1899/1971, 'Ueber Gegenstände höherer Ordnung und deren Verhältnis zur inneren Wahrnehmung', Gesamtausgabe Vol. II.

Meinong, A.: 1902/1977 *Ueber Annahmen*, Gesamtausgabe Vol. IV, Graz (2nd ed. 1910).

Meinong, A.: 1903/1969, 'Bemerkungen über den Farbenkörper und das Mischungsgesetz', Gesamtausgabe Vol. I. (BFM)
Meinong, A.: 1906/1973, 'Ueber die Erfahrungsgrundlagen unseres Wissens', Gesamtausgabe Vol. V. (UEW)
Moore, G. E.: 1959, 'Wittgenstein's Lectures in 1930—33', *Philosophical Papers*, Allen & Unwin, London.
Mulligan, K.: 1988, 'Seeing as and Assimilative Perception', *Brentano Studien*, **I**, 129—152.
Mulligan, K.: 1990, 'Criteria and Indication', *Proceedings of the 1989 Kirchberg Wittgenstein Conference*, 94—105.
Mulligan, K.: 1990a, 'Concetti formali nel pensiero di Wittgenstein e Husserl', *L'uomo, un segno* (*Proceedings of the 1989 Milan Wittgenstein Conference*), forthcoming.
Mulligan, K.: 1991, 'Wittgenstein on what we see', *Proceedings of 1989 Rome Wittgenstein conference*, forthcoming.
Mulligan, K.: 1991a, 'On the Logic and Psychology of Perception', *Proceedings of 1990 Gargnano conference on 'The Logical Form of Perceptual Reports'*.
Simons, P.: 1991, *Essays on Meinong*, Philosophia, Munich, forthcoming.
Smith, B. (ed.): 1982, *Parts and Moments. Studies in Logic and Formal Ontology*, Philosophia, Munich.
Smith, B. (ed.): 1988, *Foundations of Gestalt Theory*, Philosophia, Munich.
Stucchi, N.: 1991, 'Seeing and Thinking: Vittorio Benussi and the Graz School', in Simons, 1991.
Westphal, J.: 1977, 'The Complexity of Quality', *Philosophy*, 59.
Westphal, J.: 1987, *Colour. Some philosophical problems from Wittgenstein*, Blackwell, Oxford.
Witasek, S.: 1907, *Grundlinien der Psychologie*, Meiner, Leipzig.
Witasek, S.: 1910, *Psychologie der Raumwahrnehmung des Auges*, Winter, Heidelberg.
Wittgenstein, L.: 1968, *Philosophical Investigations*, Blackwell, Oxford. (PI)
Wittgenstein, L.: 1972, *The Blue and Brown Books*, Blackwell, Oxford. (BB)
Wittgenstein, L.: 1975, *Philosophical Remarks*, Blackwell, Oxford. (PR)
Wittgenstein, L.: 1977, *Remarks on Colour*, Blackwell, Oxford. (RC)
Wittgenstein, L.: 1980, *Remarks on the Philosophy of Psychology*, Vols I & II, Blackwell, Oxford. (RPPI, RPPII)
Wittgenstein, L.: 1981, *Zettel*, Blackwell, Oxford. (Z)
Wittgenstein, L.: 1982, *Wittgenstein's Lectures. Cambridge 1930—32*, Blackwell, Oxford. (LWL)
Wittgenstein, L.: 1982, *Last Writings on the Philosophy of Psychology*, Volume I, Blackwell, Oxford. (LWI)

TERENCE PARSONS

ATOMIC SENTENCES AS SINGULAR TERMS IN FREE LOGIC

In logic texts, it is generally assumed that verbs denote properties of, or relations among, individuals. This is in marked contrast to the common view that verbs stand for events, actions, or states. In this paper I explore the common view about verbs. This quickly turns into a revised view, according to which it is whole 'atomic' sentences (of English) that do the denoting, not just the verbs themselves. This is then most naturally formalized as if sentences of English were singular terms within a version of free logic, with the true ones successfully denoting events, and the false ones failing to denote.

INTRODUCTION

In dictionaries and grammar books we are likely to find definitions such as these:

noun: a word that stands for a person, place, or thing.

verb: a word that expresses an action, event, or state.

The fact that explanations of this sort are so ubiquitous suggests that they must contain some insight; the interesting question is whether that insight can be given a useful theoretical characterization.

The semantics of nouns, as opposed to verbs, has received a great deal of scrutiny. We now know that the way in which nouns 'stand for things' is a matter of some subtlety and complexity. A proper noun, such as 'Kim' or 'Samantha', refers to its bearer, which is indeed a 'thing', but common nouns such as 'table' or 'giraffe' do not stand for things of this sort; they stand for *kinds* of things. We do use common nouns to refer to particular things, but only when the nouns are coupled with definite articles or demonstratives: '*the* table', '*that* giraffe'. In addition, and of great importance, we also use them to quantify over things, in phrases such as 'some tables' or 'no giraffes'. A large part of the study of modern logic is taken up with the study of complex noun phrases of this sort.

Much the same is true of verbs: ordinary verbs are not names of particular actions; instead they stand for *kinds* of actions, events, or

Wolfgang Spohn et al. (eds.), Existence and Explanation, 103—113.

states, and it is a matter of some complexity to formulate a theory of how we use them to talk about particular actions and states. The thesis to be explored here is that simple sentences that have main verbs are used to refer to events and states characterized by those verbs. True sentences refer successfully; those that fail to refer successfully are false. Crudely put, 'Brutus stabbed Caesar' contains a purported reference to a stabbing of Caesar by Brutus; the sentence is true if and only if this reference is successful. The verb 'stab' makes a key contribution to the reference, but other elements of the sentence are relevant too.

To get a grip on this idea we need a way to conceptualize the structure of English sentences for semantic purposes. I assume a structure analogous to that of formal languages. In this structure we assume there are ATOMIC FORMULAS, which are combined with tenses, quantifiers, and connectives to form full sentences. They are also embedded in other constructions, such as that-clauses. The sort of structure I have in mind is displayed in the following 'generation' of the complex sentence 'Mary believes that Brutus stabbed Caesar with a knife':

ATOMIC:	x stab y with z
	↓
TENSE:	x stabbed y with z
	↓
QUANTIFICATION AND NAME-FILLING:	Somebody stabbed Caesar with a knife
	↓
EMBEDDING IN THAT-CLAUSE:	Mary believes that somebody stabbed Caesar with a knife

There is more to it than this, of course. 'Mary believes' itself has a structure that I have left unanalyzed. The sentence may also be combined with others by connectives, and this may happen either before or after quantification, which itself may happen either before or after embedding in that-clauses, and so on.[1]

The point of the exposition of structure given above is simply to isolate my topic, which is to investigate the semantics of the initial 'atomic' clauses of English from which the final structures are generated. My goal is an understanding of phrases such as 'x stab y with z'. The account that I will propose is the one that is implicit in much popular talk of language, but that is rarely taken seriously in formal work in semantics. It is the idea that such phrases refer to actions, events, or states. This idea is worth pursuing because of its intuitive naturalness, but there is also another reason: such phrases have a logical structure, and construing them as clauses referring to actions, events, and states allows an analysis of that structure.

For simplicity in this essay I will ignore state sentences, such as '*x* is under *y*' or '*x* believes that *p*'; I will concentrate instead on action or event sentences, such as '*x* stabs *y*'. Also, I will typically cite examples containing names or definite descriptions instead of variables. In the official theory, singular terms get into the game at a more advanced level. But examples using variables are intuitively awkward to process, so I will use names and descriptions instead. When you read 'Agatha sees the elephant' you should officially construe this as " '*x* sees *y*' in a context in which Agatha is assigned to '*x*' and the elephant to '*y*'."

THE BASIC THEORY

The theoretical framework I will explore stems originally from the work of Panini, sometime before 200 B.C. His account is something like the following. In a simple sentence such as 'Kim kit the tree with a knife' the verb stands for some particular action. The nouns in the sentence stand for people or things, and the sentence asserts that those things stand in certain relations to the action. For example, Kim is the agent of the action, the knife is the instrument of the action, and the tree is the object of the action. This view is endorsed independently by Plato, at least in simplified form; he held that in a simple sentence the verb denotes an action, and the noun denotes the performer of the action.[2]

This is close to the account that will be discussed in this essay, except that the account given here denies that verbs, by themselves, stand for particular actions. For one thing, the same verb 'hit' occurs with a wide variety of different agents and objects. If 'hit' by itself were understood as referring to a particular action, then this would have to be a different action in different sentences — since the actions in 'Mary hit John' and 'Sam hit Fred' are different. (E.g. one might have been violent, the other not.) The line I will investigate takes verbs to be more like common nouns than proper nouns. On this view, the verb 'hit' in 'Mary hit Fred' does not stand for a particular action (a particular hitting), it stands for a kind of action, for hittings. The clause *as a whole* refers to an action of that kind, and says that it took place; this is an action that has Mary as agent and Fred as object. A simple sentence using 'hit' refers to a hitting, and says that it took place, one using 'eat' refers to an eating, and says that it took place, and so on. The (atomic) sentence is semantically equivalent to a definite description, which refers to the event; the verb itself contributes to the description part of the definite description. If the verb occurs without modifiers, as in 'Mary saw Bill', then it is easy to mistakenly attribute to the verb the referring function that is really due to the whole clause.

THE ACCOUNT

The account can be summed up by the following principles and illustrations:

A main verb is symbolized as a predicate of events.
E.g. 'stab' is symbolized as 'Stabbing ( )', which can be read '. . . is a stabbing'.

Subjects and objects are related to the event in question by appropriate relations.
E.g. in 'Brutus stabbed Caesar' Brutus is represented as Agent, and Caesar as Object, in clauses of the form:
Agent (. . . , Brutus) Object (. . . , Caesar)
where the dots represent a place for the event variable.

The whole main clause is symbolized as a definite description.
E.g. ignoring tense, the symbolization of the sentence 'Brutus stabbed Caesar' would contain the definite description:
(ιe) [Stabbing (e) & Agent (e, Brutus) & Object (e, Caesar)].

The sentence as a whole says that the event occurs:
Occur (ιe) [Stabbing (e) & Agent (e, Brutus) & Object (e, Caesar)].

Tenses complicate the account by linking the event with an event time, and saying that this time takes place in the past, present, or future. This is achieved by quantification outside the scope of the description:
E.g. the full representation of 'Brutus stabbed Caesar' is: 'For some time t in the past, the stabbing by Brutus of Caesar at t occurred.' Or, in

symbols: $(\exists\ t)\ [t <$ now & Occur (ιe) [Stabbing (e) & Agent $(e,$ Brutus) & Object $(e,$ Caesar) & e is at t]].
For simplicity, I will ignore tenses in what follows whenever it is feasible to do so.

TROUBLE WITH FALSITY

We now have an account that appears to work as expected for true sentences. But what about false ones? Suppose I say 'Mary hit Fred', but that I am wrong. Then what does 'hit' refer to in my sentence? (More accurately, what does the *sentence* refer to?) There are two natural options here; we can say that the sentence fails to refer when it is false, or that it succeeds in referring to an action which does not occur. Either of these views might be developed into a viable account. I will explore only the former: if Mary did not hit Fred, then there is no event of Mary's hitting Fred, and the sentence 'Mary hit Fred' fails altogether to refer. This makes the view in question a candidate for analysis within free logic.

The idea that false (atomic) sentences are ones that fail to refer to actions or events also has a theoretical advantage; it allows us to analyze the notion of 'occurring' for events in terms of the familiar notion of 'existence' in free logic. That is, we can simply represent:

The stabbing of Caesar by Brutus *occurs*

by:

The stabbing of Caesar by Brutus *exists.*

Let us make the usual supposition that 'exists' ('*E!*') applied to an empty singular term yields a false sentence. Then the theory seems to work as expected. In detail, the logical form of 'Mary hit Fred' is now:

$$E!(\iota e)\ [\text{Hitting}\ (e)\ \&\ \text{Agent}\ (e, \text{Mary})\ \&\ \text{Object}\ (e, \text{Fred})].$$

If Mary never hit Fred, then the contained definite description fails to refer and the representation is false, as expected.

WHY BOTHER?

This theory looks highly artificial and convoluted. Why make clauses into names of events? What is gained besides needless complexity and a forced analogy to free logic?

Naturally, I cannot prove that the free logic approach is preferable to others, since there are well-known ways to get many of the effects of free logic within classical logic; the whole history of this issue is full of controversy. But I can address the rest of it — the issue of 'needless complexity'.

The proposed sentential forms are considerably more complicated than the traditional symbolizations found in logic texts, and one should ask whether, and why, the additional complication is necessary. The answer is that the additional structure involving events or states provides an account of various logico-linguistic phenomena. These are explored within a slightly different theoretical framework in Parsons, *Events in the Semantics of English* (MIT Press, 1990), hereafter 'Parsons (*ESE*)'. I list some of them here for reference:

The logic of modifiers: We need to account for logical relations among sentences such as that between 'Brutus stabbed Caesar in the back' and 'Brutus stabbed Caesar'; e.g. we need to explain how the former entails the latter. If verb modifiers are seen as predicates of the denoted event, then the two sentences are symbolized as:

$E!(\iota e)$ [Stabbing (e) & Agent (e, b) & Object (e, c) & In (e, back)]

and:

$E!(\iota e)$ [Stabbing (e) & Agent e, b) & Object (e, c)].

It is apparent then that the former entails the latter by ordinary predicate logic (assuming that Brutus only stabbed Caesar once; we will return to this point below).

The semantics of perception statements: We need, e.g., to capture the relation between 'Agatha saw Brutus stab Caesar' and 'Agatha saw the stabbing of Caesar by Brutus'. It is a simple matter within the framework under discussion to give these sentences related representations, since according to that framework they both contain references to the same event.

The semantics of causatives and inchoatives: Why does 'Mary will open the door' entail 'the door will open', and why does this in turn entail 'the door will be open'? (Note that in traditional logic these three uses of 'open' are symbolized as three completely independent predicates.) The framework of underlying events and states allows us to hold that the first sentence refers to an event of Mary's which causes an opening of the door; this caused event is the event that is referred to in the second sentence. Further, openings can be characterized as events which are becomings of certain states, so that the predicate of states in 'The door is open' can be used to analyze the predicate of events in 'The door opens'. (See (*ESE*) for details.)

Relations between the explicit ways in which we talk about events and the sentences that we use that do not appear to involve explicit reference to events at all. There are important relations between 'A flight over the pole by a Norwegian took place in May 1926' and 'A Norwegian flew over the pole in May 1926'. Or between the NP 'a violent destruction of a city' and the VP 'violently destroyed a city'. Related to these examples are the relation between the semantics of adjectives ('slow') and their corresponding -ly adverbs ('slowly'), and between verbs ('sing') and related common nouns ('the singing'). In fact, if we identify the meanings of the adjectives and the corresponding adverbs, and identify the meanings of the verbs and their corresponding common nouns, this framework easily yields an equivalence between the sentences cited above, and reveals the logical relations between the indicated NP and VP.

These topics, and a host of others, can be addressed by a theory of underlying events and states. There is a wealth of data in our native tongue just waiting to be explained; the hypothesis of underlying events and states can be used to explain a great deal of it. The linguistic details are pursued in (*ESE*); the present essay addresses the basic semantic issues.

A DIFFICULTY WITH THE THEORY

I turn now to a difficulty with the theory as developed so far. It has to do with the presumed uniqueness of the definite descriptions used in the sample sentences. For example, suppose that Brutus stabbed Caesar twice, and suppose further that both stabbings were simultaneous. Then the definite description that we use in the analysis of 'Brutus stabbed Caesar' seemingly fails to refer, making the analysis of the sentence false:

> $E!(\iota e)$ [Stabbing (e) & Agent (e, Brutus) & Object (e, Caesar)].

Yet this is incorrect; the English sentence is true, not false. I see two possible solutions to this difficulty. The first involves complicating the interpretation of the theory as given, and the second involves moving from definite to indefinite descriptions.

Solution # 1: Space and Time

To see how this solution works, we must first reintroduce the times that have been omitted from the symbolization for purposes of simplicity. For the past tense sentence 'Brutus stabbed Caesar' we do not have:

> $E!(\iota e)$ [Stabbing (e) & Agent (e, Brutus) & Object (e, Caesar)],

we have instead:

> $(\exists t)$ [$t <$ now & $E!(\iota e)$ [Stabbing (e) & Agent (e, Brutus) & Object (e, Caesar) & e is at t].

This now solves the problem of multiple events which are *not* simultaneous. Suppose, for example, that Brutus stabbed Caesar at several different times. Then the original simplified representation (incorrectly) comes out false, because the definite description in it is not uniquely satisfied. But the refined version comes out true: there is *at least one time* before now such that a unique stabbing of Caesar by Brutus at that time exists.

However, the problem remains to account for the problem of simultaneous stabbings. Suppose that Brutus has stabbed Caesar exactly twice, and that those stabbings were simultaneous. Then the refined analysis also (incorrectly) comes out false. The problem has not yet been solved.

I can think of one way out of this. It is apparent that space and time are similar and bound up with one another. So it might be natural to construe the 't' in the tense quantifier as ranging not over times alone, but over spatio-temporal regions. The past tense quantifier can then be read as: 'at some spatiotemporal region located wholly in the past'. This leaves the symbolism alone, and only complicates its interpretation. But this may be enough to solve the problem facing us. For there may be a past spatio-temporal region at which there is exactly one stabbing of Caesar by Brutus, even though there is another such stabbing at a contemporaneous region. This assumes, of course, that the two simultaneous stabbings of Brutus by Caesar occupy different *spatial* regions, and this is tricky, since it is not clear which spatial regions such events occupy.

To see how this might work, suppose that both Brutus and Caesar were straddling the gateway of the agora, and suppose that the two stabbings were sidearm, and that the knives went into Caesar's opposite shoulders. Then perhaps the one stabbing was in the agora, and the other was not. This would then seem to yield different locations for the two stabbings. By analogy, other simultaneous stabbings could be similarly distinguished.

There remains the problem of distinguishing distinct but simultaneous events that lack spatial locations, such as thinkings. It is not clear whether this version of the theory can survive attack-by-counterexample from this direction.[3]

Solution # 2: Indefinite Descriptions

There is another solution worth considering. This is to make the descriptions indefinite instead of definite, using a free logic variant of Hilbert's epsilon operator. The idea here is that an indefinite description denotes one of the things that the description part is true of, if anything; otherwise it fails to refer. That is, '(an *x*) [*Fx*]' refers to a thing that satisfies '*Fx*' if one or more things satisfy '*Fx*', and otherwise it fails to refer. The sentence 'Brutus stabbed Caesar' is symbolized (ignoring times) as:

> *E!*(an *e*) [Stabbing (*e*) & Agent (*e*, Brutus) & Object (*e*, Caesar)].

This has no difficulty with simultaneous stabbings, since the indefinite description succeeds in referring in such a case. If Brutus never stabbed Caesar, then the indefinite description fails to refer, and the whole sentence is false.

This theory is subject to a problem of indeterminacy: we have said that an indefinite description refers to one of the potentially many things that satisfy its description part, but we have not said which. This means that we have not fully specified the semantics. We know, of course, that whatever method is given for choosing a referent for the indefinite description in such a case will be as correct as any other, so the choice among methods is not important. However, if there is *no* method, then I don't see that we yet have a theory. And specifying a method of selection for any indefinite description that occurs in the discussion of any subject matter whatsoever is not a thing I know how to do.

This problem could probably be bypassed by an appeal to supervaluations. Namely, we assign to a sentence of the form:

$$E!(\text{an } x)\,[Fx]$$

whatever truth-value it has on all assignments to '(an *x*) [*Fx*]' of a thing that satisfies '*Fx*', if there is one. This then gives the right answer without requiring a method of selection of assigning referents to indefinite descriptions.

Putting aside the question of method selection, the theory that employs indefinite descriptions in free logic now appears to me to be equivalent to a classical logic approach in which existential quantifiers are used. This is because the anlaysis:

$$E!(\text{an } e)\,[\text{Stabbing }(e)\ \&\ \text{Agent }(e, \text{Brutus})\ \&\ \text{Object }(e, \text{Caesar})],$$

is always equivalent in truth to:

$$(\exists e)\,[\text{Stabbing }(e)\ \&\ \text{Agent }(e, \text{Brutus})\ \&\ \text{Object }(e, \text{Caesar})].$$

If this is right, then the analysis of language using indefinite descriptions in free logic is equivalent to that given in Parsons (*ESE*), and the two are equally viable.

SUMMARY

The intuition that verbs denote actions, events, or states, suitably refined and coupled with a free logic analysis of such denotation, can be developed into a revealing semantics for the atomic sentences of natural language.

Philosophy Department,
University of California,
Irvine, U.S.A.

NOTES

[1] I assume that 'name-filling' is like quantification in that it can occur at any of various stages; this accounts for the fact that names, as well as quantifiers, have 'scope'. In a logic such as Quine's version of Russell's, names of natural language are represented by quantifiers; in a free logic they are not; in either case they need to have scope. The treatment of names is well-studied and debated; I do not pursue it here.

[2] Plato says: "... we have two kinds of vocal indications of being. ... One called nouns, the other verbs. ... The indication which relates to action we may call a verb. ... And the vocal sign applied to those who perform the actions in question we call a noun." He goes on to refer to "... 'walks', 'runs', 'sleeps' and the other verbs which denote actions." Plato, *Sophist* 261D—262C.

[3] Suppose that Brutus imagines Caesar as a tyrant, and also simultaneously imagines him as a griffin. Then the sentence 'Brutus imagined Caesar' may give trouble. For the description '*e* is an imagining & Agent (*e*, Brutus) & Object (*e*, Caesar) & *e* happens at *t*' will apparently not be individuating.

ERHARD SCHEIBE

EPR-SITUATION AND BELL'S INEQUALITY

I

Following the original paper of Bell[1] the connection between the EPR-argument for the incompleteness of quantum mechanics[2] and Bell's inequality is usually established by saying that the former suggests the existence of a local hidden-variable theory for quantum mechanics whereas the latter then allows to show that such a theory does not exist.[3] More specificily the following account is given. The EPR-argument is based on two premises:

The reality criterion: If, without in any way disturbing a system I, we can predict with certainty (i.e., with probability equal to unity) the value of a physical quantity F, then there exists an element of physical reality corresponding to F.[4]

The locality condition: If the prediction for F of I can be made by measuring a quantity G of a second system II, spatially seperated from and not interacting with system I, then that prediction *can* be made without disturbing I.[5]

To get the argument started the following situation is envisaged:

EPR-situation: The total system I $\otimes$ II is in a state Φ in which every pair out of a set M of pairs (F, G) of quantities of I and II respectively is EPR-correlated.

Here F (of I) and G (of II) are said to be *EPR-correlated* in state Φ (of I $\otimes$ II) iff there is a one-to-one correlation $\alpha \leftrightarrow \beta$ between the possible values of F and G such that, if a measurement of F yields α, an immediately following measurement of G would, on account of Φ, certainly yield β, and *vice versa*.

Given an EPR-situation characterized by Φ and M we can now argue, by first using the locality axiom and subsequently the reality criterion, that quantities F of I and G of II occurring in M have 'real' values, i.e. values that are not 'created' by a measurement. In other words, there is a state hidden behind Φ in which all the quantities of I and II occurring in M have definite values. Now, although M can be very comprehensive, given Φ, it can certainly *not* contain *all* quantities

Wolfgang Spohn et al. (eds.), Existence and Explanation, 115—129.

from I or II. Ignoring this it is usually concluded that, according to EPR, there should be hidden states covering *all* quantities from I and II. And, of course, one can argue that, given any quantity F of I, *there is* always a state Φ of I $\otimes$ II and a quantity G of II such that F is EPR-correlated with G in Φ and *vice versa*. In this way, although no single EPR-situation allows the conclusion, their totality *somehow* may bring it about.

At any rate, the second part of the account — the one bringing in the Bell inequality — starts with the assumption that there are (total) hidden states and that they are local in a certain sense that has first been specified by Bell. There is, in my view, a problem connected with the manner in which *this* condition of locality, i.e. the one that is actually used in refuting the assumption of a local hidden-variable theory, comes out of the original locality condition formulated above and put into their argument by EPR. Since this is one of the major points of contact between EPR and Bell I shall comment on it in the last section of this paper. For the moment we may assume that, *if* the transformation of the original idea of locality into Bell's version of it is made, then the rest of the argument is straightforward: The assumption of a classical probabilistic and local hidden-variable theory that can reproduce all of the probabilistic predictions of quantum mechanics has the consequence that Bell's inequality would have to be generally valid also in quantum mechanics. This not being the case that assumption is refuted. Now my main concern in this paper will be to raise the following question: Since it is obvious that the assumption of an EPR-situation is vital for the EPR-part of the whole argument, what is its relation to Bell's inequality as the main tool in the Bell-part of the argument? The following is obvious: If the set M in an EPR-situation consists of at least three pairs of observables one has freedom enough to violate the inequality (in the version (3) below). But what if there are only two pairs? In this case only two of the four pairs occurring in (3) below are not strictly correlated, and this suggests the conjecture that Bell's inequality holds without any further assumption. A proof is given in the simplest case where the Hilbert spaces of the subsystems are 2-dimensional.[6]

II

To prepare the ground let us first recapitulate what Bell's inequality says and what its status in quantum mechanics is. For any self-adjoint elements A, B, A', B' of a *-algebra $\mathcal{A}$, pairwise commutable as they appear jointly in the following expression, and for any positive linear form E on $\mathcal{A}$ we define

$$(1) \quad \Delta_E(A, A', B, B') \equiv |E(AB) - E(AB')| + |E(A'B) + E(A'B')|,$$

In the physical interpretation the self-adjoint elements of $\mathcal{A}$ are observables, and E is an expectation-value function. Two main cases are to be distinguished: The *classical* case where $\mathcal{A}$ is a *-algebra of integrable functions on a classical probability space and (μ being a measure on the space)

$$(2a) \qquad E_c(A) = \int A \, \mathrm{d}\mu$$

and the *quantum mechanical* case where $\mathcal{A}$ is an irreducible *-algebra of bounded linear operators on a Hilbert-space and

$$(2b) \qquad E_q(A) = Tr(WA)$$

with a statistical operator W. If the spectra of the A, B, A', B' are absolutely bounded by 1, then in the classical case Bell's inequality

$$(3a) \qquad \Delta_\mu^c(A, A', B, B') \leqslant 2$$

and in the quantum mechanical case

$$(3b) \qquad \Delta_W^q(A, A', B, B') \leqslant 2\sqrt{2}$$

holds.[7]

The proof of (3a) is straightforward:

$$\begin{aligned}
\Delta_\mu^c(A, A', B, B') &= |E_c(A(B - B'))| + |E_c(A'(B + B'))| \\
&\leqslant E_c(|A|\,|B - B'|) + E_c(|A'|\,|B + B'|) \\
&\leqslant E_c(|B - B'|) + E_c(|B + B'|) \\
&= E_c(|B - B'| + |B + B'|) \\
&\leqslant 2
\end{aligned}$$

where the last step comes from the corresponding (trivial) inequality for the argument of E_c in the last but one line. I have given this proof because in view of the literature it seems to be the most simple and reliable proof existing. Moreover, in view of the host of 'derivations' of Bell's inequality from unclear premises that prevail in the literature[8] I want to emphasize the status of the inequality as being just a theorem of classical probability theory. As such it could have been found in the 18th century.[9] Saying this is not belittling the importance of its recent discovery which, of course, lies in the physical application. However, to be involved in physics may not dispense us from trying to argue as clearly as possible. And those 'derivations' of Bell's inequality would gain much in clearness if they were reconstructed as so many proofs from the axioms of classical probability theory together with, but strictly seperated from, considerations as to how and why these axioms come into play as the physical circumstances may require.

In the quantum mechanical case the inequality (3a) has to be replaced by the weaker statement (3b) in order to be generally valid. However, as is well known there are several more special assumptions under which

(3c) $\Delta^q_W(A, A'; B, B') \leqslant 2$

also in quantum mechanics. Since I want to add another such case it may be worthwhile to review briefly the more important cases already known.

1) (3c) holds if *all* observables A, B, A' and B' pairwise commute.[10]

Obviously the additional assumption restores the classical situation so that the result is to be expected. Accordingly, the proof (for discrete observables) is given by using the additional assumption for introducing a common orthonormal basis of eigen-vectors for A, B, A' and B'. This basis then is immediately turned into a classical probability space on which our operators reappear as real-valued functions. The rest of the proof is identical with the one given for (3a).

The next case to be mentioned is the case of a quantum mechanical system composed of two subsystems I and II. As we know from the introduction this is the case we shall resume for our main consideration. In it the four observables A, A' and B, B' are assumed to belong to the subsystems I and II respectively. More precisely we assume

(4) $A = F \otimes 1, \quad A' = F' \otimes 1, \quad B = 1 \otimes G, \quad B' = G'$

with F, F' and G, G' being observables of I and II respectively. Our general assumption about commutability is then satisfied automatically.

For the moment the special case to be mentioned is

2) (3c) holds if W is a product, i.e. if $W = U \otimes V$ where U belongs to I and V to II.

The proof is immediate if one considers that

$$Tr((U \otimes V)(F \otimes G)) = Tr(UF) \cdot Tr(VG)$$

which reduces the situation to the classical one.

Our third case shows how the classical behaviour of Δ_W^q can be hereditary:

3) If (3c) holds for statistical operators W_i then also for the convex sum $\sum_i \varrho_i W_i$ where $\rho_i \geqslant 0$ and $\sum_i \rho_i = 1$

Here the proof is the classical one *ab ovo*. From (2) and (3) it follows immediately that

4) (3c) holds if $W = \sum_i P_{\varphi_i \otimes \psi_i}$

This result[11] is remarkable if compared with the EPR-situation to which we now turn.

III

In the EPR-situation we deal not with a mixture as in 4) but with the corresponding pure state

$$(5) \qquad \Phi = \sum_i a_i \varphi_i \otimes \psi_i$$

with orthonormal bases $\{\varphi_i\}_i$ and $\{\psi_i\}_i$. In contrast to 4) for such a state (3c) does not, of course, hold without further ado. But it seems to hold if we assume that of the observables involved, i.e. of F, G, F' and G', F and G as well as F' and G' are EPR-correlated in the sense of the introduction. Let us be more explicit about the allimportant concept of EPR-correlation. It concerns observables F and G of the subsystems I and II of a compound system. We assume F and G to have a discrete spectrum so that we have spectral decompositions of the form

$$(6) \qquad F = \sum_i f_i P_i^F, \quad G = \sum_i g_i P_i^G$$

Now it may happen that with respect to a pure state Φ of the compound system I $\otimes$ II the two observables F and G are correlated in the following sense: There is a one-to-one correspondence between the possible outcomes f_i and g_i of a measurement of F and G respectively (here assumed to be established by the identity of indices) such that in Φ the probability of the result f_i for F and G_j for G of a joint measurement is 0 if $i \neq j$:

$$(7a) \qquad \|(P_i^F \otimes P_j^G)\Phi\|^2 = 0 \qquad (i \neq j)$$

This assumption is equivalent with the requirement that the *conditional* probability of f_i of F given g_i of G (or *vice versa*) is equal to 1:

$$(7b) \qquad \frac{\|(P_i^F \otimes P_i^G)\Phi\|^2}{\|(1 \otimes P_i^G)\Phi\|^2} = 1$$

So here we have the situation envisaged by EPR that the observable F of I is measured by a measurement of G of the *other* system II: Given the result g_i of the latter (7b) allows us to *predict* the result f_i for the observable F.[12]

The major conjecture now is:

Conjecture: If in the case of a compound quantum mechanical system I $\otimes$ II the observables F, F' of I are EPR-correlated with G, G' of II respectively with respect to a pure state Φ of I $\otimes$ II then (under the usual assumption concerning the spectra) Bell's inequality

$$\Delta_\Phi^q(F \otimes 1, F' \otimes 1; 1 \otimes G, 1 \otimes G') \leqslant 2$$

holds.

It may immediately be added that, as the examples given in the literature show, this conjecture cannot be strengthened by dropping one of the two correlation assumptions. I do not have a proof of this conjecture in general. But here is the sketch of a proof for the simplest case in which I and II have 2-dimensional Hilbert spaces. (This proof, therefore, covers the 'spin case' that is virtually the only one discussed in the literature.)

Without loss of generality we can assume that F and F' (and consequently G and G') do not commute. (For it they did our conclusion would follow already according to case 1) above.) Because of the EPR-correlations Φ then has two essentially different decompositions (5). Using the polar decomposition theorem it follows that in any decomposition (5) of Φ the a_i have equal absolute values.[13] From this it can easily be concluded that there is also the decomposition

$$(8) \qquad \Phi = \frac{1}{\sqrt{2}} (\varphi_1 \otimes \psi_2 - \varphi_2 \otimes \psi_1)$$

where we have re-named the ψ-vectors to achieve the familiar notation. If in this representation we choose

$$(9a) \qquad \begin{cases} F\varphi_1 = (\alpha + w)\varphi_1 + (u - iv)\varphi_2 \\ F\varphi_2 = (u + iv)\varphi_1 + (\alpha - w)\varphi_2 \end{cases}$$

and correspondingly

$$(9b) \qquad \begin{cases} G\psi_1 = (\beta + z)\psi_1 + (x - iy)\psi_2 \\ G\psi_2 = (x + iy)\psi_1 + (\beta - z)\psi_2 \end{cases}$$

then a straightforward calculation yields

$$(10a) \qquad \langle \Phi | (F \otimes G)\Phi \rangle = \alpha\beta - UX$$

where in vector form

$$(10b) \qquad U = (u, v, w), \quad X = (x, y, z), \quad UX = ux + vy + wz$$

In the next step we make essentially use of the assumption that F and G are EPR-correlated. This means that

$$(11) \qquad \Phi = a_1 \bar{\varphi}_1 \otimes \bar{\psi}_1 + a_2 \bar{\varphi}_2 \otimes \bar{\psi}_2$$

with eigen-vector bases $\{\bar{\varphi}_i\}_i$ and $\{\bar{\psi}_i\}_i$ of F and G respectively. It is not too difficult to infer from this that there is a (possibly different) decomposition but again with eigen-vector bases such that there is *one* unitary matrix u_{ik} which transforms $\{\varphi_i\}_i$ in $\{\bar{\varphi}_i\}_i$ and $\{\psi_i\}_i$ in $\{\bar{\psi}_i\}_i$. Consequently, the matrices representing F and G according to (9) commute (as do the corresponding diagonal matrices in the representation (11)). This finally immediately leads to

$$(12) \qquad U = \lambda X, U' = \lambda' X' \qquad (\lambda, \lambda' \neq 0)$$

where we have included the case of F' and G'. The inequality to be proven is thus

$$\text{(13a)}\quad \begin{aligned} &|(\alpha\beta - \lambda X^2) - (\alpha\beta' - \lambda XX')| \\ &+ |(\alpha'\beta - \lambda' XX') + (\alpha'\beta' - \lambda' X'^2| \leqslant 2 \end{aligned}$$

Now the eigenvalues of F are $\alpha \pm |U|$, and similarly for the other operators. The spectrum condition under which (13a) has to be proven therefore is

$$\text{(13b)}\quad |\alpha \pm |\lambda X||, |\alpha' \pm |\lambda' X'||, |\beta \pm |X||, |\beta' \pm |X'|| \leqslant 1.$$

The rest of the proof is dominated by case distinction which, however, can be easily reduced to the principal case in which the numbers within the $|\ldots|$ in (13a) are both non-negative. Call their sum Δ. Using the Cauchy-Schwarz inequality we easily get

$$\begin{aligned} \Delta \leqslant\ &(\alpha\beta - \lambda|X|^2) - (\alpha\beta' \mp \lambda'|X||X'|) \\ &+ (\alpha'\beta \mp \lambda'|X||X'|) + (\alpha'\beta' - \lambda'|X'|^2) \end{aligned}$$

according to whether $\lambda \gtrless \lambda'$. To simplify the notation we follow up only the case $\lambda > \lambda'$. Abbreviating the right-hand side of the last inequality by Δ_1 the reader will easily verify the identity

$$\begin{aligned} 2\Delta_2 = &\ a_+(b_- - b'_-) + a'_+(b_- + b'_-) \\ &+ a_-(b_+ - b'_+) + a'_-(b_+ + b'_+) \end{aligned}$$

where $a_\pm = \alpha \pm \lambda|X|$ and similarly in the other cases. It follows

$$\begin{aligned} 2\Delta_1 \leqslant &\ |a_+||b_- - b'_-| + |a'_+||b_- + b'_-| \\ &+ |a_-||b_+ - b'_+| + |a'_-||b_+ + b'_+|. \end{aligned}$$

Using (13b) we obtain

$$\begin{aligned} 2\Delta_1 &\leqslant |b_- - b'_-| + |b_- + b^{\pm}_-| + |b_+ - b'_+| + |b_+ + b'_+| \\ &\leqslant 4 \end{aligned}$$

where the last step again uses (13b). The final result now is immediate.

What does our conjecture, if it holds in general, mean? As was already said in the Introduction, it does in general not mean that we have to go beyond an EPR-situation in order to violate Bell's inequality: If M consists of more than two pairs of observables violation is possible in principle. It is, therefore, not necessary to conceive of the hidden states as covering more observables than occur in an EPR-situation in order to show their non-existence.[14] However, Bell's theorem cannot be extended to the lowest case: Of the four pairs of observables occurring in the inequality either none or one or two can be EPR-correlated. In the latter case violation of (3c) is not possible, and the case Bell vs. EPR was thus not the easiest.

IV

Einstein said:[15] "But on one supposition we should, in my opinion, absolutely hold fast: the real factual situation of the system II is independent of what is done with the system I, which is spatially separated from the former." Bell translated:[16] "The vital assumption is that the result B for particle II does not depend on the setting *a*, of the magnet for particle I, nor A on *b*." Was that a correct translation? I will approach the answer by starting out from the following question: Suppose we want to show the non-existence of *any* classical hidden-variables theory for quantum mechanics by using the fact that Bell's inequality is valid classically but not in quantum mechanics. How would we proceed? I think the key inference can hardly be other than the following: If the theory in question existed then, since the inequality holds classically, it would have to hold also in quantum mechanics. The question, therefore, is how to make this inference.

Again one can hardly think of any other possibility than that the equation

$$(14) \qquad E_q((F \otimes 1)(1 \otimes G)) = E_c(\sigma(F \otimes 1) \cdot \sigma(1 \otimes G))$$

is the decisive link between the two theories. Here E_c and E_q are from (2a) and (2b) respectively, and it is assumed that

(α) to every E_q (as in (2b)) there exists an E_c (as in (2a)) such that (14) for all F and G.
σ in (14) assigns to every quantum mechanical observable its classical counterpart. If we assume that

(β_1) $\quad A$ and $\sigma(A)$ have the same spectrum

the inference sought can now be made: Given E_q we have E_c with (14). Given observables F, G, F', G' with the usual spectrum premise for (3) the classical counterparts $\sigma(F)$, etc. will also satisfy this premise.

Consequently, they satisfy (3a). By (14), the F, G, F' and G' would have to satisfy (3b). Now, though already (14) looks quite nice one is tempted to replace it by

$$(14') \qquad E_q(A) = E_c(\sigma(A)),$$

(α) by

$$(\alpha') = (\alpha) \text{ with } (14') \text{ instead of } (14)$$

and to insert the additional postulate

$$(\beta_2) \qquad \sigma((F \otimes 1)(1 \otimes G)) = \sigma(F \otimes 1) \cdot \sigma(1 \otimes G)$$

from which (14) obviously follows. (β_2) is nothing but a condition of isomorphy: σ maps the observables on classical quantities in such a way that the image of the (quantum mechanical) product is the (classical) product of the images.

If at this point it were asked whether a locality condition had already been put into the argument it would be hard to find one. And yet, as shown, the argument is perfect. Accordingly, we have to set out and find locality. To this end we must take into account that, according to the assumptions introducing (2a), the classical quantities by which the quantum mechanical observables are represented via σ, are functions on a state space, say, S. But this means, by duality, that the states are functions on the algebra of quantities:

$$(15a) \qquad s(f) = f(s) \qquad (s \in S)$$

Moreover, by definition

$$(15b) \qquad (f \cdot g)(s) = f(s) \cdot g(s)$$

and therefore on account of (15a)

$$(15c) \qquad s(f \cdot g) = s(f) \cdot s(g)$$

Using σ we can now define the hidden states also as functions on the algebra of quantum mechanical observables:

$$(16) \qquad s(A) = s(\sigma(A)).$$

Now by the isomorphy (β_2) the product rule (15c) can immediately be transferred to the observables with the result

$$(\beta_2') \qquad s((F \otimes 1)(1 \otimes G)) = s(F \otimes 1) \cdot s(1 \otimes G).$$

We thus see that there is a substantial restriction on the hidden states if these are viewed as functions on the set of quantum mechanical observables: (β'_2) exactly corresponds to the classical equation (15c) and is equivalent to (β_2) if we assume

(γ) if $f(s) = g(s)$ for all $s \in S$ then $f = g$

as another obvious axiom about the classical states and quantities.

(β'_2) is almost Bell's locality assumption as it is actually used. We only need to go one last step by observing the equivalence of (β'_2) with

(β''_2) $s((F \otimes 1)(1 \otimes G))/s(1 \otimes G)$ is independent of G, and $s((F \otimes 1)(1 \otimes G))/s(F \otimes 1)$ of F.

Here independence means that the value of the first (second) quotient is the same for all $G(F)$. In order to understand the meaning of (β''_2) it is useful to observe that this condition is obviously violated by most expectation value functions. This will become even more obvious if we generalize the equivalent (β'_2) of (β''_2) to become

(N) For all commutable observables A, B: $s(A \cdot B) = s(A) \cdot s(B)$.

For expectation-value functions (N) means that they are *dispersion-free*. Now already Von Neumann had proven that *there are no* dispersion-free expectation-value functions in quantum mechanics, and this fact was interpreted by him and many followers to the effect that there is no hidden-variables theory for quantum mechanics.[17] Later on this interpretation was criticized with the argument that Von Neumann's conception of hidden states had been much too restrictive.[18] Indeed Bell's 'locality' assumption (β'_2) is much weaker than (N) because the homomorphism is required to hold only for observables A and B that belong to different physical systems. It was in this way that hidden states were made *logically* possible again even if their lifetime turned out to be very short: They soon became the victims of a *probabilistic* argument, i.e. of Bell's theorem.

I think it is fair to say that already Von Neumann's famous proof did dispose of Einstein's original idea of a hidden state.[19] For Einstein "there is such a thing as the 'real state' of a physical system — something that objectively exists independently of any observation or measurement . . .".[20] But the hidden states of the post Von Neumann era actually became dependent on measurements.

According to Bell the Von Neumann tradition "tacitly assumed that measurement of an observable must yield the same value independently of what other measurements may be made simultaneously".[21] Indeed, what can it mean that the value of an observable A in a state s *depends on another observable* B? It does not mean that A is a function of B in the usual sense. It rather means the following: First there is no dependence of A on B insofar as the value of A is uniquely determined by the state s. But secondly, there is dependence in the sense that for observables B, *commensurable with* A, the quotient $s(A \cdot B)/s(B)$ in general has different values for different B. These two features can go together on the ground of the following interpretation in terms of measurements.[22] It is assumed that every measuring apparatus has a well determined separation power in the following sense: There is a unique observable A such that, if a measurement is performed with this apparatus, there will be a unique outcome for A but for no observable finer than A. This outcome of a *direct* measurement (as it could be called) is the one that is uniquely assigned to A by any state. However, in general we can also (directly) measure observables finer than A and then *infer* the value of A from the direct measurement. But this very common practice hinges on a tacit assumption of independence: If, for instance, instead of measuring A (directly) we measure the finer observable AB then the inference in question would be impossible if $s(A \cdot B)/s(B)$ depended on B. The generalized hidden states showing just this dependence, we can 'understand' it by 1) viewing states as giving only the value of each observable that would come out if a (direct) measurement were made, and 2) giving up the idea that results of measurement have the independence described.

The foregoing considerations seem to show that Bell locality is a far cry from Einstein locality. I do not want to become indulged in questions of terminology. But it may at least be mentioned that, whereas Einstein locality being essentially the principle of action by contact, obviously is linked to space, it is hard to see any *immediate* relationship between Bell's condition and space. However, the main point is that Einstein locality has essentially to do with *interaction, presupposing,* I think, the common, classical way of describing a physical system in terms of states as quoted above. By contrast, Bell's condition concerns already the *kinematical* or *state* description of a system. It first accepts *a very uncommon* generalization of the concept of state by allowing the dependencies explained above and then mitigates the general situation for the case of systems consisting of subsystems. In accordance with the idea that the generalized states be used as hidden states of a quantum mechanical system the approach mirrors the situation brought about by the critical quantum mechanical states (5) which has also nothing to do with interactions.[23] I am afraid that physicists may be inclined to think that Bohm's 1951 hidden-variables theory illustrates Bell's theorem. But whereas the non-local interaction inherent in Bohm's quantum potential indeed violates Einstein locality[24] the theory has not yet been adjusted to the theoretical apparatus within which Bell's locality condition is meaningful.[25] Consequently, we cannot tell what it would mean that the theory is or is not 'local' in Bell's sense. Saying this is not denying that *there is* some (possibly deep) connection between the two localities. It is only to remind us that it is still waiting for being understood.

Department of Philosophy
University of Heidelberg
Heidelberg, Germany

NOTES

[1] Bell (1964).
[2] Einstein *et al.* (1935).
[3] See, for instance, the review papers Clauser/Shimony (1978) and Selleri (1988).
[4] This is essentially the famous wording of the criterion in the original EPR-paper Einstein *et al.* (1935).
[5] The locality condition is more explicit in Einstein (1948) and (1949) than in the EPR-paper.
[6] Profs. Bell, Selleri and Van Fraassen have confirmed my impression that this question has not been dealt with in the literature.
[7] (3a) is, then, the version of Bell's inequality that appears in Bell (1971). A proof of (3b) under very general assumptions is given in Summers/Werner (1987), p. 2442.
[8] Besides the papers mentioned in no. 2 see also Selleri/Tarozzi (1981).
[9] For precursors of Bell's inequality see the references in Pitowsky (1989), p. 49.
[10] I did not find this case in the literature.
[11] Selleri (1988), pp. 28f.
[12] It should be remarked that at this point the EPR-argument uses Von Neumann's projection postulate. Interestingly enough, whereas there are countless attacks on this postulate to be found in the literature I have *never* seen anybody blaming EPR for using it.

[13] Of course, this can be established also directly. The use of the polar decomposition theorem was suggested to me by Frank Artzenius.
[14] Recall that Bell actually did confine his consideration to the spin observables, see his (1964).
[15] Einstein (1949), p. 85.
[16] Bell (1964), p. 196; see also ibid. no. 2, p. 200.
[17] For a systematic presentation of this tradition see Scheibe (1981).
[18] Belinfante (1973), Bell (1966), and many others.
[19] The belittling of Von Neumann's result in Belinfante (1979), Ch. I, is unjustified in view of this function.
[20] Einstein (1953).
[21] Bell (1966), p. 451.
[22] For details of this part of the paper see Scheibe (1986).
[23] Bohr (1935), p. 700.
[24] See Bohm's own emphasis in Bohm/Hiley (1975).
[25] This is admitted in Belinfante (1973), p. 94.

REFERENCES

Belinfante, F. J.: 1973, *A Survey of Hidden-Variables Theories.* Oxford.

Bell, J.: 1964, 'On the Einstein-Podolsky-Rosen paradox', *Physics* **1**, 195—200.

Bell, J.: 1966, 'On the Problem of Hidden Variables in Quantum Mechanics', *Rev. Mod. Phys* **38**, 447—52.

Bell. J.: 1971, 'Introduction to the hidden-variables question', *Proc. Int. School of Phys. 'Enrico Fermi'*, course IL: Foundations of Quantum Mechanics, New York, 171—81.

Bohm, D. J. and Hiley, B. J.: 1975, 'On the Intuitive Understanding of Nonlocality as Implied by Quantum Theory', *Foundation of Physics* **5**, 93—109.

Bohr, N.: 1935, 'Can Quantum-Mechanical Description of Physical Reality be considered Complete?', *Phys. Rev.* **48**, 696—702.

Clauser, J. and Shimony, A.: 1978, 'Bell's theorem: experimental tests and implications', *Rep. Prog. Phys.* **41**, 1881—1927.

Einstein, A., Podolsky, B., and Rosen, N.: 1935, 'Can quantum-mechanical description of physical reality be considered complete?', *Phys. Rev.* **47**, 777—80.

Einstein, A.: 1948, 'Quantenmechanik und Wirklichkeit', *Dialectica* **2**, 320—3.

Einstein, A.: 1949, *Albert Einstein. Philosopher-scientist*, ed. by A. Schilpp, Evanston, Illinois.

Einstein, A.: 1953, 'Einleitende Bemerkungen über Grundbegriffe', in: *Louis de Broglie. Physicien et Penseur*, Paris (German Edition Hamburg 1955).

Pitowsky, I.: 1989, *Quantum Probability — Quantum Logic.* Berlin.

Scheibe, E.: 1981, 'Quantentheorie und verborgene Parameter', *Der Physikunterricht* **15**, 56—74.

Scheibe, E.: 1986, 'What kind of hidden variables are excluded by Bell's inequality?, in: *Foundations of Physics.* Papers Contrib. to the 7th Int. Congr. of Logic, Methodology and Philosophy of Science, ed. by P. Weingartner and G. Dorn, Vienna, 251—71.

Selleri, F., and Tarozzi, G. 1981, 'Quantum Mechanics, Reality, and Separability', *Riv. del Nuovo Cimento* **4**, No. 2. 1—53.

Selleri, F.: 1988, 'History of the Einstein-Podolsky-Rosen paradox', in: *Quantum mechanics versus local realism*, ed. by F. Selleri, New York, 1—61.

Summers, St. J. and Werner, R.: 1987, 'Bell's Inequalities and Quantum Field Theory', *J. Math. Phys.* **28**, 2440—56.

PETER M. SIMONS

ON BEING SPREAD OUT IN TIME: TEMPORAL PARTS AND THE PROBLEM OF CHANGE

1. TWO VIEWS ON CHANGE

What is change? There are two opposed views: the *substratum* view, which goes back to Aristotle, and the *replacement* view, which goes back at least to Hume. The *substratum* view is that change consists in an object (the substratum) having first one attribute, then another attribute contrary to the first. Here is Aristotle's formulation:

> Since change is from one contrary (or an intermediate) to another contrary, there must be something underlying which changes into the contrary state; for the contraries themselves do not change. Further, since the contraries pass away into each other, there must be some third thing in addition to them which persists throughout the change; this thing is the material.[1]

According to the replacement view, change consists not in one thing's successively receiving contraries but in different successive things having the contraries, these things being conceived as temporal parts or phases of a whole which is extended in time. Hume says

> That action of the imagination, by which we consider the uninterrupted and invariable object, and that by which we reflect on the succession of related objects, are almost the same to the feeling, nor is there much effort of thought requir'd in the latter case than in the former. The relation facilitates the transition of the mind from one object to another, and renders its passage as smooth as if it contemplated one continu'd object. This resemblance is the cause of the confusion and mistake, and makes us substitute the notion of identity, instead of that of related objects.[2]

In *On the Plurality of Worlds* David Lewis argues that what I call the substratum view cannot explain the possibility of change, so that we are compelled to recognize that ordinary individuals like houses, trees and people have temporal parts, and are therefore spread out in time. In this paper I examine Lewis's argument and defend the substratum view against his criticisms. I do not think the reasons advanced for dispensing with continuants have shown themselves to be sufficiently weighty to warrant abandoning them, while the costs of seriously trying to do without them have been underestimated. My previous arguments to this effect[3] did not take account of Lewis's criticisms of continuants,

Wolfgang Spohn et al. (eds.), Existence and Explanation, 131–147.

and as this has been held against me[4] I feel beholden to confront the objections. In any case, Lewis's case is made strongly, so if a successful defence can be made of continuants in the face of his attack they emerge with strengthened ontological credentials. The controversy raises a number of difficult issues about the nature of predication and the connections between language and reality which bear on the interests of the dedicatee of this volume.

Each view of change has two aspects: an ontological aspect concerning the nature of the objects of which change is predicated, and a linguistic aspect concerning the nature of predications about such objects. The substratum view regards the subjects of change as objects which, while they *persist* in time, are not extended in time, that is, do not have temporal parts. Such objects are *continuants*: they *endure* over time. The replacement view takes the objects of which change is predicated to be temporally extended: they are *occurrents* and *perdure* over time.[5] Typical predications about continuants, ascribing them an attribute (a property, or a relation to something else), are not semantically definite in truth-value unless there is some reference to a time, whether explicit or implicit, context-free or indexical, or else some device tantamount to quantification over times. Thus, to say of a certain apple *A* that it is red is not to express a proposition with a truth-value unless it is clear from the context when the apple is red, e.g. because the present tense is used. By contrast, when on the replacement view it is said that a certain temporal phase or slice of an apple is red, this statement is not semantically indefinite: the phase or slice has its time of existence built into it, and it is thus true or false without need of further specification that it is red.

I shall consider both aspects of the controversy, trying also to pick out what I think is correct about Lewis's view. The discussion will mainly concern intrinsic change, change concerning a single continuant; change *among* continuants (such as relative motion of bodies) raising no new problems for my purposes.

2. LEWIS ON TEMPORARY INTRINSICS

On the Plurality of Worlds is mainly about modality, and the argument against continuants on pp. 202—4 is an analogue of modal arguments against individuals being wholly in different possible worlds. Lewis rejects the idea that an individual can exist as a whole at different times:

the only way an individual can straddle times is to have temporal parts which wholly exist at different times, so a transtemporal individual is a mereological sum of instantaneous individuals. Lewis's argument is that the alternatives to this view are unacceptable:

> The principle and decisive objection against endurance, as an account of the persistence of ordinary things like people or puddles, is the problem of temporary intrinsics. Persisting things change their intrinsic properties. For instance shape: when I sit, I have a bent shape; when I stand, I have a straightened shape. Both shapes are temporary intrinsic properties; I have them only some of the time. How is such change possible? I know of only three solutions. . . .
>
> First solution: contrary to what we might think, shapes are not genuine intrinsic properties. They are disguised relations, which an enduring thing may bear to times. . . . This is simply incredible . . . If we know what shape is, we know that it is a property, not a relation.
>
> Second solution: the only intrinsic properties of a thing are those it has at the present moment. . . . This is a solution that rejects endurance; because it rejects persistence altogether. And it is even less credible than the first solution. No man, unless it be at the moment of his execution, believes that he has no future; still less does anyone believe that he has no past.
>
> Third solution: the different shapes, and the different temporary intrinsics generally, belong to different things. Endurance is to be rejected in favour of perdurance. We perdure; we are made up of temporal parts, and our temporal intrinsics are properties of these parts, wherein they differ from one another. There is no problem at all about how different things can differ in their intrinsic properties.[6]

Most of Lewis's critics agree with him about the first two views, and look for some fourth way out. The main problem is to prevent such solutions from either missing the point or collapsing into one of those rejected.

3. TEMPORAL MODIFICATION: SEMANTIC ALTERNATIVES

An intrinsic property is one which concerns the way a thing is, irrespective of its relationships to anything else.[7] Being an uncle is an extrinsic or relational property, involving my brother or sister having a child: becoming an uncle is a relational or Cambridge change rather than a real change.[8] Not wishing to prejudice the analysis of the metaphysics of change in favour of universals, I consider *predication*. In the simplest possible kind of predication a predicate, which is a linguistic expression or part of one, is said or predicated of a subject, which is the entity denoted by the subject *term* of the predication.[9] A predication which, if true, is so in virtue of the way the subject is in itself, may be called an intrinsic predication.

Intrinsic predications cannot be discerned from logical form alone, as the uncle example shows. Counting noun phrases in a sentence is only the beginning. Predications involving adjectives admitting of degree for example involve implicit comparison with a standard or reference class, as becomes clear when this shifts. The same person who is tall in Tokyo is short in Amsterdam without suffering irreversible Lorentz contraction on the flight between.

The troublesome predications about continuants start out with the apparently intrinsic form *A is F*, but since the continuant changes in respect of *F* the time must somehow be taken into account. For brevity I shall not distinguish tense from adverbial or other temporal modification, subsuming all forms under the term *tense*. One is led to consider the form *A is F at t*. How is it to be analysed? What are its basic components and how are they structured? In particular, what, if anything, does the temporal modifier *at t* modify? There are surprisingly many suggestions:

(1) *A* is-*F*-at *t*

This is Lewis's first option. The predication expressed a relation, the *is-F-at* relation, between the subject *A* and a time *t*.

(2) *A* is *F*-at-*t*

Here temporal modification affects the predicate term.

(3) *A* is-at-*t* *F*

Here not the property but instantiation of the property is tensed, as marked by the tensed copula. This is a popular 'fourth option' in reply to Lewis.[10] It is not always clearly distinguished from

(4) (*A* is *F*) at *t*

Here subject and property compose a state of affairs which obtains at some times and not at others, this being expressed by the temporal modification. The whole predication rather than any part of it is tensed. Likewise easily confused with previous alternatives is

(5) *A* ((is *F*) at *t*)

Here tense modifies a predicate which is intrinsically predicative but untensed, the resulting tensed predicate then being said of the subject.

(6) *A*-at-*t* is *F*

This is Lewis's favoured option. The subject of the intrinsic predication is a time-slice of *A*, or rather, since this suggests the wrong order of priority, *A* is an aggregate of successive momentary objects, each of which is a temporal part of the whole.

Each of these suggested parsings connects with a suggested ontological analysis of what makes the predication true. It would however be wrong to say that we can read the ontology off the parsing or vice versa. Jonathan Lowe has suggested that Lewis's problems stem in part from running together a semantic and a metaphysical problem.[11] However if we aspire to the idea that a perspicuous representation of what makes a sentence true denotes or quantifies over only entities we accept in our ontology, then such a persicuous representation will have consequences for the ontology, but it is *not* likely to be given by a parsing of what is there in the superficial form of the sentence, nor is it guaranteed to be recoverable by meaning analysis. What makes a sentence true is generally opaque to linguistic investigation.[12] So let us suppose each of the six cases represents a perspicuous representation of what makes a simple predication true. What are the costs and benefits of the alternatives?

Ad (1): Lewis spurns this option: "If we know what shape is, we know it is a property, not a relation."[13] The ridicule is slightly misplaced: not *straight* or *is straight* but *is straight at* expresses the relation. The form (1) embodies the idea that the relation *is-F-at* is the prime constituent, from which *is F* is somehow derivative, perhaps in the way that being an uncle or being tall are derivative from being uncle of or being tall in comparison with. I think this is a strange analysis, but it is not quite so plainly idiotic as Lewis, by his choice of words, makes it seem. Intrinsic predications do not after all wear their characters on their sleeves.

Ad (2): The meaning of this is obscure. Being straight is not intrinsic, being straight at *t* is. But how is being straight related to being straight at *t*? Again presumably by derelativization or derivation, which leaves this option little better off than the first. The relation of *A* to this tensed predicate depends on how the copula 'is' is understood. Does the 'is' represent a timeless relation, or one which obtains only when the subject has the predicate? If the second, then the modification of the predicate appears superfluous and the case reduces to (3). It is also difficult to see how the analysis works for intransitive verb predications, which do not have a separate copula and complement, since the modification applies only to the complement and leaves the copula untouched.

Ad (3): This associates tense with the copula, and is thus subject to a similar objection to the last. Given that in copulaless sentences tense directly modifies the main verb, it tends to run into (5). But assuming that talk about modifying the copula is pseudo-grammatical talk about the relation of instantiation obtaining between subject and property, it is then unacceptable to nominalists. If, like Lewis or myself, you have little time for universals and therefore for the idea of instantiation as a relation (or 'tie', or 'nexus') between a universal and an individual, this option is of little use.

Ad (4): This option is as attractive — or unattractive — as the notion of a tensed state of affairs which obtains at some times and not others. What is its status at times at which it does not obtain? Does it exist then or not? If it does, it is very like Lewis's alternative possible worlds which exist but are not actual, except that it is partial and obtains temporarily. If obtaining is a kind of property, we seem simply to have moved Lewis's problem to states of affairs as subjects: instead of *A is F at t* we have *A's being F obtains at t*, and we are no wiser. If obtaining is just existing for states of affairs, then while we do not have Lewis's problem, we have exchanged Lewis's problem for the problem of temporary existence, which is at least as difficult, though it is one which any theory, including Lewis's, has to face.

Ad (5): This very much fits the way tense works in natural languages, both with copula + complement predicates and intransitive verbs. It is distinguished from (4) in that a predication takes place only after the temporal modification, not before it. Metaphysically the main problem is to see how to distinguish it from options (1) and (3), and especially to see how it escapes Lewis's charge of making something intrinsic extrinsic, since presumably in this case too the predicate *is F* is not in itself already ripe for predication. However it has a strong advantage over all the other options but (4), especially over the options (1)—(3), which it resembles more closely than it resembles (6). In (1)—(3) if we omit the time *t* the *at* dangles and we need a special account of how either an unmarked time (the present) or an existential quantification (some time) leads to the loss of the preposition. The all-purpose possessive *of* is naturally lost in derelativization, being there in the first place because the relative noun used (e.g. *uncle*) does not itself dangle a place. In (5) the *at* is part of the modification (now = at the present), so its ready deletion is explained. Because in (1)—(3) the preposition is built into the analysis its deletion is not easy to explain. By contrast (5) puts the prepositional phrase as a whole *at t* in an exposed and elidable position. The difference between a tensed and an untensed predication is also readily shown: in an untensed predication like *7 is prime* the predicate *is prime* is untensed (here of course I mean semantically untensed). Unlike e.g. (3), the analysis (5) makes no metaphysical commitments of its own about the nature of predication: it inherits whatever account we want to give of predication in general, whether involving universals or not. The main task is to say what temporal modification amounts to, and defend it against Lewis's attack on option (1).

Ad (6): This option is of course not new, but it goes back further than one might think, being found explicitly in the writings of Bolzano in the early nineteenth century and Leśniewski in the early twentieth. Later it has been associated with Carnap, Quine, Smart and others. It stands out from all the other alternatives even on a linguistic level. Since Aristotle it has been assumed — on the basis of the obvious phenomena of ordinary languages — that the tense of a predication is intimately connected with its verb, its main functor. Options (1)—(3) and (5) respect this in their different ways, option (4) takes the tense to affect the predication as a whole. Option (6) alone distributes tense to the subject. Here the deletion of *at t* has the effect not of modifying the predicate or the predication but of giving us a new subject to deal with, one which is *more extensive*. In the case of many tensed relational predications, like *Socrates was shorter than Plato in 400 B.C.*, or *John gave Mary the book on December 25th 1955*, the modification must distribute to the several subjects, so we are talking about simultaneously existing temporal parts of each of them.

This brief survey suggests that purely from a point of view of linguistic naturalness and metaphysical neutrality option (5) has the edge. The naturalness claim needs to be supported by more extensive empirical linguistic investigations. For now I shall assume for the sake of argument that the claim can be supported. The question then is whether option (5) exacts an excessive price, as Lewis claims.

4. TEMPORAL MODIFICATION: METAPHYSICAL ALTERNATIVES

One reason for looking askance at the idea that apparently intrinsic characteristics are 'really' relations to times is that one might have scruples about taking times seriously as entities, *a fortiori* as entities to which one is related. Following Leibniz, one might take relationships of precedence, simultaneity and so on among events as being ontologically prior, times being some kind of logical constructions from these. One may recall that such relations are themselves not absolute in special relativity, but depend on the frame of reference. This of course has repercussions also for Lewis's theory of temporal parts, because it means that which parts of a spatially extended persisting object are temporal parts, i.e. at right angles to the time axis, likewise depends on the frame of reference. This is not an objection, since anyone who accepts that some objects, namely occurrents, have temporal parts, must draw the same conclusion.

Where the defender of option (5) has to concede a point to Lewis is that by accepting temporal modification one is admitting that the predicate *is F* is in some sense not absolute. Of course the form of words *is F* may occur in a predication without further ado, either because the present is understood or because a generalization is involved. But each of these supposes the priority of the tensed case – assuming we reject with Lewis the idea that only the present is real.

Many predications about occurrents are not significantly tensed. An explosion *E* is not an explosion *at a time*, even though it occurs at a time. It just is an explosion. The *is* is timeless, and expresses a pure classification. E's being an explosion is intrinsic, but since there is no time *at which* it is an explosion, and no time *at which* it is not an explosion, its being an explosion is neither a temporary nor a permanent intrinsic. One might *legislate* that it is an explosion only when it is occurring, or that it is an explosion for ever and always, but each is artificial, the first especially so. What is it when not occurring? Something other than an explosion? But what? Presumably then when not occurring it is nothing? The oddity of all these questions all point to it being without sense to tie *E*'s being an explosion to a time. Occurrents are however not alone in being subject to predications not significantly tensed. Pure classificatory and essential predications about continuants share this feature. John is a human being, an animal and self-identical, again, just is these things, not at some time. The sentence *John is a human being* can also be understood as tensed, being true only when John exists. But the tensed sense presupposes the untensed, since it must be determinate *what* exists at a certain time.

Both occurrents and continuants enter into significantly tensed propositions however, though neither is concerned with change in the items concerned. When a battle is raging it is true to say it is going on, happening, taking place or occurring, whereas before or after it such a statement will be false. Likewise when John is alive it is true to say that he exists or is alive, but false to say it before he comes to be or after he dies. The *exists* and *is* here are significantly tensed. Admittedly we find it barbarous to say of occurrents, as distinct from continuants, that they exist rather than happen or occur, but this is a relatively superficial matter. Both continuants and occurrents exist, or, if you prefer, *are* at certain times and — provided they are not everlasting — not at others. In view of our scepticism about times as entities, it would be inadvisable to say they bear an existing-relation to times. Rather, they exist simultaneously with, before and after others. The differences between them for those who deny that continuants have temporal parts is that whenever an occurrent exists some temporal part of it exists, and different temporal parts of it exist at different times when it exists, whereas continuants simply have no temporal parts to exist at different times.

Now consider tensed predication for occurrents. A party which starts off quietly may get noisier. Its doing so consists in its early parts being quiet and its later parts being noisy. This is a variation of (let us suppose) intrinsic properties among its parts, a successive variation. Continuants may vary in the properties of their parts too: some white cats have one blue eye and one orange eye. But this is not variation among temporal parts and is not successive, so it is no candidate for change. Now consider a leaf which is first green and then turns red. For Lewis this is just like the party: earlier parts are green, later parts are red. It is variation of intrinsic properties among parts. For me, it is not variation among parts. Because being green and being red are incompatible properties, they cannot claim the same subject. Lewis, by spreading the subject out over time, gives the properties different subjects, temporal parts of the leaf. I cannot do that, so I must accept that because both green and red claim the same subject, they cannot do so simultaneously. That does not make being green a relation to a time, or relative to a time. John may be tall relative to Japanese and short relative to Dutchmen, the leaf is not green relative to June and red relative to October: it is green in June and red in October. *Being green* is an intrinsic property which an object has — at a time, i.e. simultaneously with some things, before others etc. The *at*, to repeat, does not signify a relation. It is precisely *because* continuants lack temporal parts that some of their intrinsic properties, namely those they might gain or lose, being theirs alone and not on secondment from temporal parts, are not theirs whenever they exist.

There is no way a continuant theorist can go further the meet Lewis's objection to solution (1) without abandoning his principled distinction. He or she must accept that some properties are just had, while others are had at a time.

I indicated above that I was unhappy with solution (3), which prescribes a metaphysics of universals. So how would I represent perspicuously what makes a simple predication true? I consider that — with a couple of exceptions — genuinely monadic predications about subjects are true in virtue of some individual feature of the subject, some dependent item variously known as an individual accident, property instance, moment, trope, particularized quality and so on. This moment m is a continuant. Being rigidly dependent on its bearer A, it cannot exist unless its bearer exists simultaneously. The mere existence of this moment m suffices to make a number of propositions true. Firstly, that m exists at that time, secondly, that it is of its natural *infima species* M and of all the genera to which M is subordinate. Because it is dependent on A it follows that at this time t there exists some M which is dependent on A, and it is this which finally makes it true that A is F at t, the predicate F being related to M roughly as 'green' is to greenness. The analysis is

(7) For some m: m is an M and m is dependent on A and m (exists (at t))

The leaf is green in June because some greenness exists in June and is dependent on the leaf. The only significantly tensed bit of this is the part saying when the greenness exists. If the leaf's colour changes in June while yet remaining green, the explanation is that a succession of greennesses qualify or inhere in the leaf which are numerically distinct but all of the genus greenness (note that each will be of some particular species of greenness).

So like Lewis I think that in any real change something comes into existence and something goes out of existence. Real change indeed involves replacement of one moment by another. But thes are not temporal parts of their bearers, and they are themselves continuants, albeit of shorter life than their bearer. Even if the concrete entity were a bundle or swarm of moments, it would not thereby gain temporal parts. If a substance is a swarm of moments, they are its parts, but temporary rather than temporal parts. In any case, I doubt whether a concrete thing is a bundle of moments.

Nevertheless whenever a subject has a property, i.e. when it has a moment of the appropriate kind, there is a putative occurrent to be considered, namely the state of its being thus, e.g. the state of the leaf's being green. The specification of this state as one of being green strongly suggests that the state is epiphenomenal: the state exists just because the leaf is green and continues to be so over a period. The existence of the state does not account for the leaf's being green, it is the other way around. And simply continuing to be green in explained in terms of the moment of greenness continuing to exist. The state gains temporal parts simply in virtue of the lapse of time, itself not a cange.

When change takes place, i.e. when moments are replaced by others, the change is itself an occurrent, an event. Events are somehow more credible as entities than states. For one thing, as Ramsey and Davidson have pointed out, many statements are happily interpreted as involving quantification over events. Similar quantification over states, while it has points in its favour, is not as well supported by the linguistic evidence.[14] Events like explosions are so salient in their own right that it appears incredible they should be logical constructions from sequences of moments, which in continuous change exist only instantaneously. It is the moments which look like the abstractions in such cases. Nevertheless what makes events the kinds of events they are has to do in most cases with the way in which their participant continuants change their properties and relations over time. Take twenty-two men, a field, two goals and a football, and get them running around kicking the ball in the right way and you have a football (soccer) game without needing to add any further ingredient. We have terms for events which mark them out in their boundaries and which tend to pick out salient, well-delimited occurrences. The difference between the salient and the humdrum is partly a matter of interest, not an ontological divide. Where occurrences go on in a way which is not well-delimited, we tend to speak of processes: the earth's circling the sun for instance. Here again the process is plausibly seen as *nothing more than* the steady change of spatial relations between the two bodies, a relation being a species or genus of relational accidents.

Apart from the rather obviously epiphenomenal nature of states, it is not easy to argue for the priority of accidents or moments over occurrents or vice versa, so I shall leave the question open. Perhaps we cannot have the one category without the other. Note that this is not the same as saying that continuants in general are constructions from occurrents. That position is more radical, and more resistable. But a complete *reduction* of occurrents to continuants seems to me to be almost as problematic. If something changes, is the change nothing, no entity at all? If events just are changes in or among continuants, they may presuppose continuants, moments perhaps amongst them, but the change, the replacing, is not unreal.

If change is as Aristotle, not Hume sees it, as alteration of a substratum, then occurrents do not change: their successive temporal parts vary. Alteration *is* on this view one thing's having first one property and then the very same thing's having another property. Occurrents on the other hand do not have temporary properties: they have temporal parts which have different properties. For example, bodies, which are continuants, move about: occurrents do not.[15] Different parts of occurrents occur at different places, but the occurrent as a whole does not move about, since, being spread out over time, it is never there as a whole to move about. So if people and other bodies were spread out over time, they would not change or move about. The motion of a body would be indistinguishable from the apparent motion of a body when a film is run. The following cases would also be conceptually indistinguishable: (1) two homogeneous cubes move towards each other until they meet, (2) at the leading edge of each cube a new thin layer comes into being while at the trailing edge a layer of the same thickness ceases to be. A continuant theorist can distinguish these cases, only the first involving translation of a body, but a temporal part theorist cannot.

5. MISSING PARTS?

In his efforts to show the substratum theory is wrong, Lewis is occasionally rhetorically less than fair to the opposition. To those who do not understand how people and houses and leaves can have temporal parts, he says,

At any rate, *I* have the concept of a temporal part; and for some while I shall be addressing only those of you who share it.[16]

To this is added a footnote,

I attempt to explain it to others in *Philosophical Papers*, volume I, pages 76—7. But I have no great hopes, since any competent philosopher who does not understand something will take care not to understand anything else whereby it might be explained.[17]

Maybe Lewis had to suffer some annoying opponents, but it is unnecessarily cynical to suggest that any philosopher who disagrees and has trouble understanding his view is being deliberately obtuse. However, I do understand the concept of a temporal part, since I know what a temporal part of an occurrent is. And I understand what it is for persons to have temporal parts: it is for them to be occurrents. So I understand perfectly what Lewis is claiming: I just think it is wrong. Persons, leaves, trees and houses do not have temporal parts.

The disagreement is major. I think that persons, etc. not only happen to lack tempral parts: they lack them of metaphysical necessity. So in the sense that no person can ever have a temporal part, I do fall in the camp of those for whom the concept of temporal part does and can not apply to continuants. Let us consider another argument that Lewis brings against this. It is that he, accepting that continuants have temporal parts, can make distinctions and understand propositions which those who do not accept his position cannot. We are poorer in conceptual resources. This is his reason for saying we are not entitled to claim he is wrong: our concept of *part* is different.

They are like the people — fictional I hope — who say that the whole of the long road is in their little village, for not one single lane of it is missing. Meaning less than others do by 'part', since they omit parts cut crosswise, they also mean less than others do by 'whole'. They say the 'whole' road is in the village; by which they mean that every 'part' is; but by that, they mean that every part cut lengthwise is. Divide the road into its least lengthwise parts; they cannot even raise the question whether those are in the village wholly or only partly. For that is a question about crosswise parts, and the concept of a crosswise part is what they lack.[15]

Leaving aside the questionable comparison of defenders of continuants with rural primitives, there is an interesting point here. Suppose there really are villagers with a road running into their village from the east and stopping there, like this:

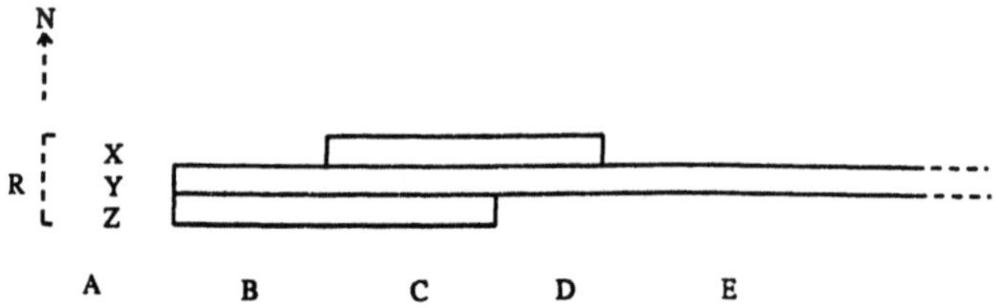

The village extends west to east from *A* to *E*, where the letters represent *longitudes*, i.e. all positions on lines running north to south. *We* say that the road has no part at longitude *A*, does have parts of different width at *B—E*, and has parts to the east beyond the village. The villagers distinguish as parts only those things we might call *lanes*, such as *X, Y* and *Z*. The road is a three-lane highway at *C*, but before reaching the eastern boundary of the village it dwindles to a one-lane track. The villagers recognize as parts of lanes only thinner lanes, not segments cut north-south. Are they conceptually impoverished? Not if they have some substitute for our talk about parts cut crosswise (north-south). It is reasonable to suppose they might have. Some predications will be east-west indexical, or longitudinally modified. The predication *R is three lanes wide* is true at *C* (and a short way to either side of it) but false at the other letters. *X is part of R* is true at *C* and *D* but false at *A, B* and *E. R exists* is false at *A*, true elsewhere.

The parallel with the controversy between substratum and replacement theorists is obvious. We, who accept crosswise parts, will say that some parts of the road have more lanes than others. The *have* is so to speak placeless. The villagers on the other hand insist that what they call *R* 'changes' along its length, in that how it is at one longitude is other than how it is at another. This is not too hard for us to understand, as we often talk this way, using temporal terminology for spatial variation, especially when thinking about a particular route from one place to the next. Roads, rivers and other roughly linear items are often said to change as we go along them, to get narrower or wider, etc.

The comparison is useful, because it helps one to understand Lewis's impatience. From our standpoint, it looks as though the villagers just simply do not recognize something we do, namely segments of the road cut north-south. True, they have a form of talk which allows them to get by, but they miss out on something: their mereology of roads is poorer than ours. But is *R* a road? Are we and they talking about the same thing? I think not. Roads have crosswise parts: *R* cannot. *R* is something of a sort we are not familiar with and for which we have no word. The villagers' ontology is not poorer than ours: it is different. If we say they simply fail to see or understand that *R* has crosswise parts, we are assuming that *R* is a road in our sense.

Suppose a geologist friend of ours declares himself a *depthist*. He insists that we, who look at the surface of the ground in terms of what is on it, miss out a whole dimension of parts of the surface. We are ignorant flatlanders. Where we say part of the earth's surface covers a part of the earth's crust which is five miles deep, the depthist insists that the surface itself is five miles deep just there. Where we say that the same piece of surface covers clay, granite and gneiss, and that these are therefore at different layers beneath the surface, the depthist insists that the surface has different depthy parts, each one of which *is* clay, granite and gneiss. I think we should rightly insist that the depthist is not talking about what we understand as surfaces, even if he carries on using our word. He is alienating the word.

The analogy between this and what I claim Lewis does in the case of continuants breaks down because we don't say the surface is wholly present at each depth. But if we take a solid cone going down from a surface area into the centre of the earth and call the entity which is wholly present at each depth a *bearing* then we have a one-one correlation between surfaces and bearings and cones. The whole bearing is present at each depth. Predicates of a bearing will sometimes be relativized to depth: is clay, is gneiss, etc. The bearing is to the cone as continuants are to Lewis's occurrents.

The cases of *R* and the road, bearing and cone, continuant and occurrent are similar, allowing for the difference between spatial and temporal extension. The lesson to be drawn is Carnapian tolerance: Lewis is not wrong to talk of temporal parts of something corresponding to me, but he is wrong to talk of temporal parts of *me*. The problem is what this something is, and how it stands to me, as it is a problem what *R* is and how it stands to the road. The correct account of Lewis's enterprise is not that he discerns parts that we miss, but that he wishes to replace the conceptual scheme of continuants having properties at times by one of occurrents with temporal parts having different properties. I have nothing against this, provided it is stated clearly that this is what is going on, and a serious attempt is made to work out the details, not simply riding on the back of our conceptual scheme. As the examples suggest, the ontologies, while describing the same spatio-temporal regions, conceptualize the content of those regions in radically different ways, and there is no easy transition from one ontology to another, no mechanical translation from one idiom to another. In particular the predicates true of the objects of Lewis's ontology cannot just be ours, or ours after a simple operation has been carried out on them. I have a vested interest in finding out about this four-dimensional entity which is so intimately related to me and my life. I do not think it *is* my life, because it has to be rich enough in content to somehow comprehend both me and my life, and it is not then a *life*, which is a life *of* a continuant. I turn a somersault: it is an event in my life. I am no part of it and it is no part of me. It is somehow submerged within the new entity, just as I am. But how? Given that the new entity and I share rather few properties, it is hard to say.[19]

Department of Philosophy,
University of Salzburg,
Salzburg, Austria.

NOTES

[1] *Metaphysics* XII, 1069b 3ff.
[2] *A Treatise of Human Nature*, Bk I Part IV §6 (Hume (1978): 253—4).
[3] Simons (1987), §3.4.
[4] Noonan (1988), 639.
[5] The continuant/occurrent terminology is due to C. D. Broad; the enduring/perduring terminology, used by Lewis, is due to Mark Johnston.
[6] Lewis (1986), 203—4.
[7] Lewis (1988), 65.
[8] Geach (1972), 321—2.
[9] I shall not be so free as to consider sentences with non-denoting subject terms. While this leads me away from garden paths well trodden by Joe Lambert, my excuse is that this problem is tough enough already.
[10] For information on who has held options (2)—(4), cf. footnotes 9 and 12 to Haslanger (1989); she holds (3). To option (4) one should add Graeme Forbes: cf. his (1987), 155.
[11] Lowe (1988).
[12] Cf. Mulligan, Simons and Smith (1984).

[13] Lewis (1986), 204.
[14] Cf. Parsons (1988), 24—29.
[15] Cf. Dretske (1967).
[16] Lewis (1986), 203.
[17] Ibid. In the passage mentioned from the *Philosophical Papers* Lewis claims that temporal parts of people are very like people, only shorter in duration: they walk, talk, etc. This is mistaken, because walking and talking, even for a short time, all involve change, which it is the point of the move to a fourth dimension to explain. Whenever what we call a change takes place, one entity replaces another. So the temporal parts have to be either instantaneous or if extended only extended so long that no change occurs in that time. Lewis is prepared to accept instantaneous person-parts, as well he must.
[18] Ibid.
[19] Thanks to Harold Noonan for challenging me to answer Lewis, to Jonathan Lowe and Sally Haslanger for sending me their helpful papers, and to members of the Philosophy Department of King's College London for stimulating discussion.

REFERENCES

Dretske, F.: 1967, 'Can Events Move?' *Mind* **76**, 479—92.

Forbes, G.: 1987, 'Is There a Problem about Persistence?' (Symposium with Mark Johnston), *The Aristototelian Society*, Supplementary Vol. **66**, 137—55.

Geach, P. T.: 1972, *Logic Matters*. Blackwell, Oxford.

Haslanger, S.: 1989, 'Persistence, Change, and Explanation', *Philosophical Studies* **56**, 1—28.

Hume, D.: 1978, *A Treatise of Human Nature*, ed. P. H. Nidditch, Clarendon Press, Oxford.

Lewis, D. K.: 1983, *Philosophical Papers*, vol. 1, Blackwell, Oxford.

Lewis, D. K.: 1986, *On the Plurality of Worlds*, Blackwell, Oxford.

Lewis, D. K.: 1988, 'Rearrangement of Particles: A Reply to Lowe', *Analysis* **48**, 65—72.

Lowe, E. J.: 1988, 'The Problems of Intrinsic Change: Rejoinder to Lewis', *Analysis* **48**, 72—77.

Mulligan K., Simons, P. M. and Smith, B.: 1984, 'Truth-Makers", *Philosophy and Phenomenological Research* **44**, 278—321.

Noonan, H.: 1988, 'Review of Simons 1987', *Mind* **97**, 638—40.

Parsons, T.: 1988, 'Underlying states in the Semantical Analysis of English', *Proceedings of the Aristotelian Society* **87**, 13—30.

Simons, P. M.: 1987, *Parts. A Study in Ontology*, Clarendon Press, Oxford.

BRIAN SKYRMS

STABILITY AND CHANCE[1]

> "The probability of an event is then relative to our degree of ignorance; but one can still think that it has an objective value in a certain sense"
> Bruno de Finetti
> *Probabilismo* Sec. 9

In *Probabilismo* (1931) de Finetti sketched a subjectivist treatment of the discourse of objective chance. Possible chances were to be taken as degrees of belief conditional on a specification of the chance set up, and the individuation of the chance set up was to be chosen with a certain kind of stability in mind. These leading ideas can be given a natural development using the notions of probability conditional on a partition or sigma algebra, and using the ergodic decomposition of invariant measures. The resulting theory casts some light on the nature of principles relating degree-of-belief and chance.

1. DE FINETTI'S THEORY OF OBJECTIVE PROBABILITY

For de Finetti a belief in the reality of objective probability is a form of superstition. This is perfectly consistent with an attempt to give some meaning to the usual talk of objective probability, and many subjectivists have had something to say along these lines. In most of de Finetti's corpus, the notion of objective probability is treated with simple disdain. But in *Probabilismo*, we find a sketch of a modest theory of objective probability as a comprehensible and sometimes useful manner of speaking. In what sense can one legitimately think about objective probability according to de Finetti? Immediately following the passage in the epigraph, he explains: "It is possible to think that someone who knows a well-determined group of circumstances and does not know the rest logically ought to evaluate the probabilities, at least of certain events, in a well-determined way." So, for example, if you knew that a die was physically symmetrical and that it will be thrown hard against a board covered with little rubber cones, etc. and knew nothing more relevant,

Wolfgang Spohn et al. (eds.), Existence and Explanation, 149–163.

you might well take the six faces as representing equiprobable outcomes. And if you were in doubt about the symmetry of the die, you might speak of being in doubt about the 'true chance'. Although you speak as if your doubt were about some hypostatized propensity, it is in fact a doubt about quite ordinary physical properties.

In a slightly idealized version of this picture, you have a finite partition of your probability space whose elements represent ways in which 'a well-determined group of circumstances' may be. Your personal conception of chance can be rendered as probability conditional on this partition. In any world in member P_i of the partition, the chance of outcome O is Pr $(O \mid P_i)$. If one is uncertain about the true member of the partition, one may be said to be uncertain about the chances. From this point of view 'chance' or 'objective probability' is simply a mathematical construction, and it is one which explains the conventional way in which one uses 'chances'. Thus, it is true by definition that probability is the expectation of chance. The idea can be given more generality by taking chances as probability conditional on a sigma-algebra, but for expository purposes we will — for the moment — talk only about finite partitions.

Since the mathematics works as well for any finite partition, this subjectivist line must face the question as to how the appropriate partition is to be chosen. We may notice with de Finetti (sec. 10) that while the subjectivist is pondering this question the frequentist had better worry about the problem of the appropriate reference class and the propensity theorist had better think about the principles of individuation for chance set-ups. One obvious idea, that de Finetti considers and rejects in paragraph 9, is to use causal relationships in determining the appropriate partition. Shouldn't everything causally relevant to the outcome be specified by members of the partition used to define its chance? de Finetti rejects this line of thought neither because it would make too many chances zero or one nor because the causal relations may be unknown, but rather for the more radical reason that causality itself has only a probabilistic meaning. For de Finetti, using causality to define chances would be circular.

Is the choice of an objectifying partition then to be made on purely pragmatic grounds, or is there some further principle that can be brought into play? de Finetti returns to the question of objective probability in section 28 of probabilismo. He discuses an example which — in one form or another — has played a role in the debate between subjectivists and objectivists throughout the history of probability theory:

An idea which — according to the usual conceptions — springs from the point of view that probability has an objective meaning, is that the valuation of probability can have a *greater* value in some cases than in others. For example, drawing a white ball from an urn that containing half white and half black balls has (on the usual hypothesis, which we presuppose, of equal subjective probability for the different balls) probability = 1/2; if the urn contains balls of the two colors in the proportion 1 to 2, but I don't know if the white or the black balls are twice the other, and the two hypotheses are equally likely, the probability is still 1/2. . . . But in the first case, when we *known* that half of the balls are white, the statement that the probability = 1/2 seems to have greater value. Naturally it will not be a *more objective* value, given that there is nothing objective in this; the subjective value, since it expresses that I am eqully uncertain between the white and black balls, does not express anything more than this in the other cases; but we do feel a difference between the cases, and we have too strong an intuitive feeling to think it is a senseless and artificial metaphysical prejudice. How shall we resolve this apparent contradiction?

Similar examples are discussed by John Stuart Mill in Chapter 18 of the first edition of *A System of Logic*, where he takes them to constitute a fatal objection to the epistemic interpretation of probability advocated by Laplace and by Sir Karl Popper,[2] who draws a similar moral. (By the 2nd edition Mill had — under the influence of his friend Sir John Hershel — completely withdrawn the criticisms.)[3] Closely related issues are inconclusively discussed by Keynes in Chapter 6 of *A Treatise on Probability* under the heading 'weight of arguments'. We know that de Finetti had read Keynes, and the term 'value' in the foregoing translation of de Finetti is closely related to Keynes' 'weight', so we may perhaps see de Finetti as giving an answer to the question that Keynes found so puzzling:

. . . it suffices to analyze how our state of mind differs in the two cases. In what does it differ? It differs in what we might call its stability. The *actual* state of mind, considered *in itself*, is really the same, and analyze it as we may we shall find no difference. But our state of mind differs if we enlarge the analysis to include our state of mind in relation to other events, from which it might or might not be independent and by which, consequently, it will or will not be modified if they occur, or if we learn of their occurrence.

Stability of degree of belief is the crucial property, and it is to be interpreted as independence; that is, as invariance under conditioning. The extent of invariance is dramatically different in the two different cases of the urn:

In case the composition is unknown, my state of mind can be influenced by learning the outcome of the preceding draws, by more or less reliable news and rumors that I might pick up about the way the urn was filled, by the visual impression I might have by glancing inside it, and by other circumstances which could give me information or hints that might diminish my doubts about the to me unknown comoposition. Whereas if the composition is known, no such circumstances can make me know it better.

Putting together what de Finetti has to say in sections 9 and 28 we may say that a subjectivist can make sense out of objective-probability-talk by construing objective probability as probability conditional on a partition (or sigma algebra) where the appropriate partition has a certain robustness property relative to the agents probabilities — probabilities conditional on members of the partition are invariant under conditioning on a wide range of possible additional information. Using these ideas, de Finetti goes on to show in section 29, how a conventional use of statistical models makes sense as a practical approximative method.

Stability is the guide by which a subjectivist is to carve her probability space at the joints. But, as de Finetti is quick to point out, stability is always relative to the wide range of possible additional information . . . "because there are always circumstances that I don't know, knowledge that would modify my state of mind." It follows that according to this treatment, some 'objective probabilities' are more 'objective' than others in that they have broader invariance properties. Thus it might be argued that probabilities for sampling from an actual urn of known composition are less "objective" than the probabilities of emission in radioactive decay. There is therefore not one sense of chance, but rather a whole family of different senses of chance, which depend on the kind of possible additional information considered.

In the following sections we will consider how these leading ideas can be developed.

2. TRUTH CONDITIONS FOR CHANCE

Suppose one wishes to give truth conditions for statements of the form: 'The chance of s is x.' Ideally, one would have a specification of the chance of s in every possible situation. That is to say, one would have a specification of a function from possible situations, w, to real numbers (which represent the chance of s in that situation). If chance so defined, is to behave the normal way in relation to subjective probability, the chance-function should at the very least, obey the principles that *the chance set-up determines the chances* and that *probability is the expectation of chance*. These formal desiderata, and some modest extensions of them, are met by the notions of probability conditional on a partition and, more generally, probability conditional on a sigma-field.

Let us presuppose a probability space, $\langle W, S, pr \rangle$. W is a set of situations specified finely enough for our purposes. (If they are maximally specific we can call them possible worlds.) S is a boolean sigma-algebra of propositions specified as measurable subsets of W. pr is a degree of belief probability measure defined on S which – contrary to de Finetti – we take to be countably additive here. Initially, let us suppose that there are a finite or countable number of ways that the determinants of the chance of a proposition, s, might be set up. This gives us a finite or countable partition, SET, of the space of possible situations according to the way the determinants of chance of s are set up in a situation. Then for any chance setup, SET_i, with non-zero probability we define the value of the function $f(w)$ for each situation, w, in SET_i as equal to the value of the conditional probability $pr(s \mid \mathrm{SET}_i)$.

$$f(w) = pr(s \mid \mathrm{SET}_i) \text{ for all } w \text{ in } \mathrm{SET}_i$$

If some member of our partition has zero probability, then the requisite conditional probability is not well-defined, but the way in which we extend our partial function to a full function will have no effect on the expectation. In this case we can choose any value for our function provided we keep s constant for all situations, w, in that member of the partition. Any function which meets these conditions is called a *version* of the probability of s conditional on the partition SET. These versions of probability of s conditional on SET form a family of functions, each of which specifies truth conditions for "The chance of s is x" in such a way that: (i) the value of this function is constant over members of the partition – the chance setup determines the chances (ii) the probability of s is the expectation of the chance of s and (iii) that probability is the expectation of chance continues to hold conditional on the possible chance setups (which have positive probability).

Property (iii) of probability conditional on a partition can, more or less, be thought of as follows. If one were to learn which chance set up obtained and nothing more, then ones probability of s would be equal to the chance of s in that chance set up. This property is closely related to a much discussed principle relating probability and chance:

PRINCIPLE M: $Pr(s \mid \text{chance of } s = x)$
$= x$ (where defined)

In fact it is easy to show that in this setting we have *M* as well.

Proposition: The definition of chance as probability conditional on a partition satisfies *M*.

Proof: By definition, the proposition "chance of $s = x$" is equal to the union of a countable number of elements, SET_i, of the partition such that $pr(s \mid \text{SET}_i) = x$ together perhaps with some elements of the partition which have zero probability and have the function defining chance arbitrarily set therein to x. If $Pr(s \mid \text{chance of } s = x)$ is well defined, then the union and thus at least one of its members has positive probability. Consider the countable union of all members, SET_j, of the partition which have positive probability, and which satisfy "Chance of $s = x$". Call this proposition *T*. By a well known property of countably additive probability measures, $pr(s \mid T)$ must be a countable average of the values of $pr(s \mid \text{SET}_j)$ for the $\text{SET}_j s$ in *T*. But for each SET_j in *T*, $Pr(s \mid \text{SET}_j) = x$, so $Pr(s \mid T) = \text{x}$. Now the proposition "chance of $s = x$" is either just *T* or the union of *T* with a proposition of zero probability. In either case $pr(s \mid \text{chance of } s = x) = pr(s \mid T) = x$.

Note: Principle *M* here does depend on countable additivity of subjective probability — a principle that de Finetti opposes. If one only has finite additivity, the proof goes through only for finite partitions.

The foregoing account shows how one can get well-behaved truth conditions for chance when the possible chance set ups form a finite or countable partition of one's probability space. But in many cases of interest we may have to consider an uncountable infinity of possible chance set ups. Here the typical case is one in which each possible chance set up has probability zero, so the elementary definition of conditional probability leaves probability conditional on a chance set up undefined. Nevertheless, many of the virtues of the foregoing construction can be retained in this case by defining chance as probability conditional on a sigma algebra.

The way to think about it is to think of the specification of a chance set up in a situation, *w*, as being made by specifying for each element, *ch*, of a sigma algebra, *CH*, [a sub-sigma algebra of *S*] whether *w* is in *ch* or not. Probability conditional on a partition becomes a special case taking *CH* to be the sigma algebra generated by the partition SET. But the sigma algebra used to specify chances need not come from such a partition. How are we to use an appropriate sigma field to set truth conditions for "The chance of $s = x$"? Consider the measure, m, defined on the sigma algebra CH by:

$$m(ch) = pr(s \cap ch) \text{ (for each } ch \in CH)$$

Application of the Radon-Nikodym theorem on $\langle W, CH \rangle$ assures the existence of a function f, measurable with respect to CH and integrable with respect to pr such that:

$$pr(s \cap ch) = \int_{ch} f\,dpr \text{ (for each } ch \in CH)$$

As before the function so specified is not unique. There is a family of such functions, any two of which disagree only on a set of probability zero. Each function represents a version of the probability of s conditional on the sigma algebra CH. If we take such a function, f, as defining chance according to:

"The chance of s is X is true in w iff $f(w) = x$

then we get the chances only depending on the propositions in the sigma algebra CH, and taking ch as the universal proposition we get that probability is equal to the expectation of chance:

$$pr(s) = \int_{w} f\,dpr$$

We also get, as a consequence of this definition, a form of principle M proposed in Skyrms (1980) Appendix 2, and adopted by Gaifman (1986). Let $[a, b]$ be a closed interval such that the probability that the chance of s is in it is positive. then, if chance is probability conditional on a sub sigma algebra, we have:[4]

PRINCIPLE M′ $pr(s \mid \text{The chance of } s \text{ is in } [a, b])$ is in $[a, b]$

The foregoing construction gives us, for a fixed proposition, s, a reasonable notion of probability conditional on the sigma-algebra CH, and thus a specification of the chance of s in each possible situation. But do we therefore have a specification of chance in general? That is, can we choose versions of "the chance of s", "the chance of t", etc. so that putting them together we have a probability measure, chance (●), for every situation? In the general abstract setting in which we have been working, the answer is no. We can exclude the kinds of mathematical pathology that make trouble by working in a *standard borel space*. That is a space which is Borel-isomorphic to a borel subset of a complete separable metric space. We will assume henceforth that we are working on a standard Borel space. Then, corresponding to our CHANCE sigma algebra, there is a chance Markov kernel — a two place function, $CH(w, s)$ such that for each situation w in W, $CH(w, \bullet)$ is a probability measure and for each proposition s in S, $CH(\bullet, s)$ is a version of the probability of s conditional on the sigma-algebra CHANCE.

Probability conditional on a partition and conditional on a sigma-algebra are standard mathematical tools which assure the existence of a subjectivistic semantics for statements of chance which meets certain formal requirements. As de Finetti points out, the semantics is highly relative. It is relative to a subjective probability space, and to an appropriate partition or sigma algebra. De Finetti suggests that *stability* considerations provide some guidance in identifying a useful sense of objective probability. We proceed to investigate how stability considerations can enter into the identification of the appropriate partition or sigma algebra.

3. STABILITY, RESILIENCY AND ERGODICITY

De Finetti defined stability as invariance under conditioning. Where one does not have absolute stability, one might still have approximate stability — a small range of movement under conditioning. If the range of movement of the probability of s under conditioning on some member of a specified set of propositions, T, is ε, call $1\text{-}\varepsilon$ the approximate stability of s with respect to T. Taken with the idea that stability of subjective probability is what gives laws of nature their apparent necessity, I wrote a book, *Causal Necessity*, investigating this notion of approximate stability, which I called 'resiliency'. I found that this notion was important in a variety of contexts. Here are three examples: Probabilistic relevance must be taken as highly resilient if it is to be taken as indicating causal relevance. Universal generalizations which are vacuously confirmed may have high probability but low resiliency, in which case they are not plausibly well-confirmed *qua* law of nature. High probability as a criterion of acceptance leads to the lottery paradox —

the set of sentences accepted may be inconsistent – no matter how high the requirement is set. But a requirement of appropriately resilient high probability guarantees consistency. I also had the same intuition that de Finetti expresses in *Probabilismo*, that stability is the key to a responsive subjectivistic treatment of what seem to be objective chances. The treatment of *Causal Necessity* was limited, however, by being mostly based on finite languages and by giving only an intuitive discussion of the sets relative to which resiliency is to be evaluated.

These two shortcomings can be remedied by using an analysis based on general ergodic theory. Ergodic theory was initially conceived as a tool for statistical mechanics but grew into a powerful abstract mathematical theory suitable for more general use. For our purposes it can be interpreted as a mathematical theory of stability of probability in the setting of countably additive probability theory.[5]

We will assume that we have our subjective probabilities defined on a standard Borel space, $\langle W, S, pr \rangle$. We pursue the individuation of the chance set up aong a slightly indirect route. We ask what counts as an instanciation of the same chance set up. Perhaps the time or place or other things can be different, but the things that determine the chances will be the same. We will take as an answer a measure preserving transformation: a mapping of W into itself which preserves the probabilistic structure. More precisely, we require of the transformation, T, that for any proposition, s:

(1) $T^{-1}s$ is a proposition (a measurable set) and

(2) $pr(T^{-1}s) = pr(s)$

If these conditions are fulfilled, the transformation T is said to be measure preserving with respect to the probability measure pr, or – shifting emphasis – the probability measure pr is said to be invariant with respect to the transformation T.

Invariance with respect to a fixed transformation, T, is a property of probabilities that is preserved under mixing. Since, in general, our subjective probabilities will be a mixture of what we take to be the possible chance probabilities, transformations which do not alter the possible chance structures should be measure preserving with respect to the subjective probability measures. The set of probability measures which are invariant with respect to T form a simplex whose extreme points correspond to the chances with respect to T. We may think of these extreme points s as the possible physical probabilities corresponding to the individuation of the chance set up implicit in T.

We might think about the individuation of the chance set up more directly. We take the propositions relevant to specifying the chance set up as the sets which are invariant with respect to T. A set, s, will be said to be is invariant with respect to T if the probability of the symmetric difference between s and its inverse image, $T^{-1}s$, is zero, It follows from this definition that the invariant sets form a sigma-algebra. We might think of the chances as specified in a situation, w, by saying for every invariant set, s, whether w is in s or not. That is, we could think of chance as probability conditional in the subsigma-algebra of invariant sets. Here we will have a Markov kernel specifying the chance probabilities in each situation, w.

Or we might try to specify the chance probabilities along the lines suggested by de Finetti's discussion of *stability*. We have adequately described the chance set up if no additional description will change the probability measure. A possible physical probability measure, ch, should such that there is no invariant set, s, such that conditioning on s would change ch. This means that for any invariant, set, s, $ch(s) = 0$ in which case conditioning on s is undefined or $ch(s) = 1$, in which case conditioning on $ch(s)$ does not change anything. Taking this together with previous discussion we might say that a possible chance measure is one satisfying two conditions:

1. ch is invariant with respect to T.
2. If s is an invariant set, $ch(s) = 0$ or $ch(s) = 1$.

If these conditions are met, ch is said to be *ergodic* (or indecomposable) with respect to T. The ergodic measures will be taken as the possible chance probabilities.

These three ways of specifying the chance probability relative to a transformation, T, all come to the same thing. The ergodic measures are the extreme points of the simplex of probability measures invariant with respect to T and they are probabilities conditional on the sub-sigma-algebra of invariant sets.

The foregoing continues to hold true when we consider not just one measure-preserving transformation, but rather a finite or countable group of measure preserving transformations.[6] The most general notion of symmetry that we have is that of a group of transformations which preserve structure. Measure preserving transformations preserve probabilistic structure. Ergodic theory thus shows us in a very general way how a symmetry in an agent's degrees of belief can be used to determine her individuation of a chance set-up and thus to determine her conception of chance.

Two agents can share the same conception of chance while having very different subjective probabilities. In this case the differing subjective probabilities can be represented as different mixtures of the same objective probabilities. Such agents will share beliefs about repetitions of an experiment almost surely generating relative frequencies which converge to the chances – via the pointwise ergodic theorem – and will share the same conception of the repetition of an experiment. The evidence of relative frequencies can be brought to bear on their initial disagreement. In this case the relativism of the subjective account of beliefs about chance is not so radical as one might at first suppose.

On the other hand, two agents might have different conceptions of the repetition of an experiments, as de Finetti forcefully points out in a later essay.[7]

> What are sometimes called *repetitions* or *trials* of the same event are for us so many distinct events. In general, they will have common or symmetric characteristics which make it natural to attribute equal probabilities to them, but there is no a priori reason which prevents us in principle from attributing to the events E1 . . . En distinct and absolutely arbitrary probabilities p1 . . . pn. In principle there is no difference for us between this case and that of n events which exhibit no similarities; the similarity which suggests the name "trials of the same event" (we would say: "of the same phenomenon") is not intrinsic . . .

If one wishes to see how drastic such a disagreement could be, the famous surreal examples of Nelson Goodman (1955) are a good place to start. For subjectivists the individuation of the chance set-up is at basis relative to an individual agent's detrees of belief, and that relativism is not totally removed, but only moderated by the analysis using the ergodic decomposition of invariant measures. As I said at the beginning of this essay, this problem is not only a problem for subjectivists. It is a problem for frequentists when they are called upon to choose the relevant reference class or reference sequence for an application of probability. It is perhaps the fundamental problem for propensity theorists. If they can do no better in solving it, then their theories take on — at least in application — the subjective and relative character emphasized by de Finetti.

Department of Philosophy,
Unrversity of California,
Irvine, U.S.A.

NOTES

[1] This paper was first read at a symposium on de Finetti's Probabilismo at the Pacific Division Meetings of the American Philosophical Association in March of 1990.

[2] See the discussion of the paradox of ideal evidence in Appendix *ix of Popper (1959).

[3] For an account of Hershel's persuation of Mill see Strong (1978).

[4] For a proof see the Appendix by Peter Vanderschraaf.

[5] As briefly reported in Chapter 3 of my *Pragmatics and Empiricism.*

[6] And for some tractable cases of uncountable groups. See Farrell (1962) Varadarajan (1963) and Dynkin (1978).

[7] de Finetti (1938).

REFERENCES

Billingsley, P. 1965 *Ergodic Theory and Information*, New York.

de Finetti, B.: 1931 Probabilismo *Logos* **14**, 163—219. Translated by M. C. Di Maio, M. C. Galavotti and R. C. Jeffrey as Probabilism, *Erkenntnis* 1989 1—55.

de Finetti, B.: 1938, Sur la condition d'equivalence partielle *Actualités scientifiques et industrielles, No. 739*, Hermann & Cie. Translated by P. Benacerraf and R. Jeffrey as On the condition of partial exchangeability in *Studies in Inductive Logic and Probability* vol. II R. C. Jeffrey (ed.) University of California Press, Berkeley, 193—205.

Dynkin, E. B.: 1978, Sufficient Statistics and Extreme Points *Annals of Probability* **6**, 705—730.

Farrell, R. H.: 1962, Representation of Invariant Measures *Illinois Journal of Mathematics* **6**, 447—467.

Gaifman, H.: 1986, A Theory of Higher Order Probabilities in *Theoretical Aspects of Reasoning About Knowledge* ed. J. Y. Halpern, Morgan Kaufmann Los Altos, California 275—292.

Goodman, N.: 1955 *Fact, Fiction and Forecast*, Harvard University Press, Cambridge, Mass.

Keynes, J. M.: 1921, *A Treatise on Probability*, London.

Mill, J. S. 1974, *A System of Logic Ratiocinative and Inductive*, University of Toronto Press. Toronto Contains versions of first edition of 1843 and second edition of 1846.

Popper, K. R.: 1959, John Stuart Mill, John Hershel, and the 'Probability of Causes' *PSA 1978* **1**, Philosophy of Science Association East Lansing, Michican, 31–41.
Skyrms, B.: 1980 *Causal Necessity*, Yale University Press, New Haven.
Skyrms, B.: 1984, *Pragmatics and Empiricism*, Yale University Press, New Haven.
Weyl, H.: 1952, *Symmetry*, Princeton University Press, Princeton.
Varadarajan, V. S.: 1963, 'Groups of Automorphisms of Borel Spaces', *Transactions of the American Mathematical Society* **109**, 191–220.

APPENDIX BY PETER VANDERSCHRAAF

A PROOF OF PRINCIPLE M′

Let $\langle \Omega, G, Pr \rangle$ be a probability space, let F be a sub σ-algebra of G, and let A be an event in G. We define the *chance of A with respect to F, ch(A)*, as the random variable

$$ch(A) = Pr(A \mid F).$$

This characterization of chance enables us to give a proof of principle M′.

THEOREM: Given an interval $[a, b] \subset [0, 1]$ with $Pr(ch(A) \in [a, b]) > 0$, then

$$(\mathrm{M}') \qquad Pr(A \mid ch(A) \in [a, b]) \in (a, b)$$

Proof: Let C be the set of points $\omega \in \Omega$ such that if $\omega \in C$ then $ch(A)(\omega) \in [a, b]$. By hypothesis, $Pr(C) > 0$. Now

$$Pr(A \cap C) = \int_C Pr(A \mid F)dPr = \int_C ch(A)dPr \; [1]$$

by the properties of conditional probability defined on a σ-algebra. Dividing both sides of [1] by $Pr(C)$, we obtain

$$Pr_C(A) = \frac{Pr(A \cap C)}{Pr(C)} = \frac{1}{Pr(C)} \int_C ch(A)dPr$$

$$= \int_\Omega ch(A)dPr_C \; [2.0]$$

$$= \int_C ch(A)dPr_C \; [2]$$

where Pr_C denotes the conditional probability measure $Pr(\cdot \mid C) = Pr(\cdot \mid ch(A) \in [a, b])$. We note that line [2] follows from line [2.0] because: i) $\Omega = C \cup C^c$ and $C \cap C^c = 0$, ii) the integral is additive over disjoint sets, and iii) C^c is a Pr_C-null set.

Now since for $\omega \in C$ we have $a \leqslant ch(A)(\omega) \leqslant b$, we obtain

$$a = a \int_C dPr_C = \int_C adPr_C \leqslant \int_C ch(A)dPr_C \text{ [monotonicity of integration]}$$

$$\leqslant \int_C bdPr_C \text{ [monotonicity of integration]}$$

$$= b \int_C dPr_C = b,$$

which coupled with [2] gives us $Pr(A \mid ch(A) \in [a, b]) = Pr_C(A) \in [a, b]$. □

We note that this proof of principle M′ is valid whether or not the range of $ch(A)$ is a finite or countable set of points.

WOLFGANG SPOHN

A REASON FOR EXPLANATION: EXPLANATIONS PROVIDE STABLE REASONS

1. INTRODUCTION*

Why ask 'Why?' Whence our drive for explanation? This is a bewildering question because it is hard to see what an answer might look like. I well remember having learnt in undergraduate courses that explanation is the supreme goal of science. So who would dare ask for more? Some fortunately did.[1] One prominent answer is that (scientific) explanation yields (scientific) understanding; and surely, we want to understand things. It is this answer which this paper is about.

When I first heard of this answer from Karel Lambert as being seriously discussed, it struck me as utterly tautological; and when arguing against it in Lambert (1988, 1991) he seemed to argue for a contradiction. However, there is one, and only one, way of rendering this answer sensible and sensibly doubtable: namely by giving independent characterizations of '(scientific) explanation' and of '(scientific) understanding' and checking how they relate. This is what Lambert (1988, 1991) did, thus recovering the full worth of the answer. But it is not what is usually done. Quite often the correctness of the answer has been presupposed, and ideas about what understanding might consist in have been built into the characterization of explanation.[2] But then the answer helps only to explicate, not to justify explanation.

Lambert (1991) concluded that the fact that an answer to a why-question is an explanation is neither sufficient nor necessary for it to yield understanding. I want to advance an argument in favour of the contrary conclusion. It needs a double stage-setting (sections 2 and 4) and has two steps (section 3 and 5). Section 2 mainly presents a general theory of non-probabilistic induction. This is the basis for section 3: a partial account of deterministic causation (which copies the probabilistic account I have given in (1983) and (1990a)) and a straightforward extension thereof to a partial account of causal explanation. Section 4 works up to some coherentistic principles in terms of the given theory of induction which involve what I shall call ultimately stable reasons. The notion of an ultimately stable reason cannot pretend

Wolfgang Spohn et al. (eds.), Existence and Explanation, 165—196.

to catch much of the rich notion of understanding; but, as section 5 explains, it fits well the characterizations of understanding which have been given in this context and may thus serve as a substitute.[3] Section 5 finally proves a formal equivalence of causal explanations and ultimately stable reasons under some restrictions which require several comments. Since the epistemological relevance of ultimately stable reasons unfolds in a coherentistic picture of truth, this equivalence construes the search for explanation as the search for truth.

2. INDUCTION AND CAUSATION

David Hume was the first to argue for an essential connection between induction and causation, so forcefully in fact that it has not ceased since to be in the focus of philosophical discussion. Indeed, for Hume induction and causation were virtually the same:

Although Hume himself struggled with the characterization of belief — believing, he said, is having ideas accompanied by a peculiar feeling of vivacity and firmness[4] —, he has an elaborate theory of belief formation. Impressions as the most lively and forceful of all perceptions are the paradigms and the basis of belief; all other empirical beliefs are gained from them by inductive extension. How? Hume held that induction proceeds just by inferring causes from effects and vice versa, i.e. via causal inferences which sufficiently, though not completely, transfer the impressions' vivacity and firmness so characteristic of belief.[5] The realm of empirical belief therefore consists of nothing but causal inferences from impressions (and their recollections).

Induction thus seems to reduce to causation. But one may as well view the matter the other way around. Hume defines causation, taken as what he calls a natural relation, as precedence, contiguity and association, i.e. transfer of liveliness and firmness, the marks of belief.[6] Thus, if A precedes B and is contiguous to it, A is a cause of B if and only if B may be inductively inferred from A. This shows that induction and causation are in effect interdefinable for Hume.[7]

The imperfections of Hume's account are well known. It is certainly wrong to say that A is a cause of B if and only if B may be inductively inferred from A, even if A and B satisfy the other conditions; symptoms of later events are clear counter-examples. And if one gives up this equivalence, it is doubtful that causal inferences exhaust inductive inferences. However, I believe that such imperfections do not defeat

Hume's fundamental insight into the essential connection between induction and causation; the task is to get it straight.

Since the ways of induction seem multifarious, it is implausible that induction should be definable in causal terms. Thus one part of this task, the one discussed in the rest of this section, must be to provide a general account of induction independent of causation. The other part, taken up in the next section, is then to say how causation relates to this account.

Concerning the first part, the first point to note is that induction and belief revision are one and the same topic: The input of an inductive scheme consists of all the information directly received, and it tells what to believe according to the input. The input of a scheme of belief revision consists of an old epistemic state and a new piece of information, and it yields a new epistemic state. Thus, a scheme of belief revision may be immediately inferred from an inductive scheme, and the latter follows from the former plus an initial epistemic state to start from. This congruence may not always have been clear because induction and revision have met different interests. Belief revisionists explicitly searched only for rationality constraints on belief, whereas the longer-standing discussion of induction tended to search for *the* correct inductive scheme, thereby presupposing, or perhaps only hoping, that there is just one such scheme possibly even independent of the initial epistemic state. History taught us, I think, that this presupposition, or hope, is misguided.[8] Therefore, the two fields have merged by now, and general accounts of induction may best be found by looking at accounts of belief revision.

Within the representation of epistemic states as (subjective) probability measures, belief revision is a rich and lively topic.[9] However, Instead want to turn to a much less familiar representation of epistemic states. One essential weakness of the probabilistic representation is that it can hardly account for *plain* belief which simply holds propositions to be true or false or neither; this is the moral of the well-known lottery paradox.[10] Therefore I dismiss probability because I want to focus on plain belief — for several reasons: First, it is of intrinsic interest to examine the structure of inductive schemes for plain belief. Secondly, if induction and causation are indeed essentially connected, then, presumably, subjective probabilities are related to probabilistic causation, whereas (sufficient and/or necessary) deterministic causation relates to plain belief; and it is the latter kind of causation I am concerned with.

Thirdly, the probabilistic counterparts of some of the assertions in the final section hold only under more restrictive conditions. Fourthly, and perhaps most importantly, subjective probabilities cannot be true or false; truth attaches only to plain belief; thus an important part of my argument will only work for plain belief.

Strangely enough, induction and revision with respect to plain belief is a much more experimental and less established field. Shackle's functions of potential surprise, Rescher's plausibility indexing, and Cohen's inductive probability[11] are pioneering contributions, and revision of plain belief has been most thoroughly studied by Gärdenfors and his coauthors.[12] In (1988) and (1990b) I have proposed a slight variant of these epistemic representations which has the advantage that it allows of generally and iteratedly applicable revision rules and thus in effect of a full account of induction for plain belief. Its basic concept is easily introduced:

Throughout, Ω is to be a set of *possible worlds* (as philosophers say without necessarily being so serious about it as is, e.g., David Lewis) or a sample space (as probability theorists prefer to say), i.e. just an exhaustive set of mutually exclusive possibilities; elements of Ω will be denoted by α, υ, ω, etc. As usual, *propositions* are represented by subsets of Ω, denoted by A, B, C, D, E, etc. The basic concept is then given by

DEFINITION 1. κ is a *natural conditional function*[13] (a *NCF*) iff κ is a function from Ω into the set of natural numbers[14] such that $\kappa^{-1}(0) \neq \emptyset$. A NCF κ is extended to propositions by defining $\kappa(A) = \min\{\kappa(\omega) \mid \omega \in A\}$ for each $A \neq \emptyset$ and $\kappa(\emptyset) = \infty$.[15]

A NCF κ is to be interpreted as a *grading of disbelief*. If $\kappa(\omega) = 0$, then ω is not disbelieved, i.e. ω might be the actual world according to κ. Because not every world can be denied to be the actual one, Definition 1 requires that $\kappa(\omega) = 0$ for some $\omega \in \Omega$. If $\kappa(\omega) = n > 0$, then ω is disbelieved with degree n. A proposition is then assigned the minimal degree of disbelief of its members.[16] Thus, if $\kappa(A) = n > 0$, then A is disbelieved with degree n. And if $\kappa(A) = 0$, then A is not disbelieved, i.e. A might be true according to κ. $\kappa(A) = 0$ does not mean that A is believed according to κ. Belief in A is rather expressed by disbelief in $\overline{A}$[17], i.e. by $\kappa(\overline{A}) > 0$, i.e. $\kappa^{-1}(0) \subseteq A$. Thus, $\kappa^{-1}(0)$ determines what is plainly believed according to κ.

Two simple properties of NCFs should be noted: *the law of negation* that for each proposition A either $\kappa(A) = 0$ or $\kappa(\overline{A}) = 0$ or both, and *the law of disjunction* that for all propositions A and B, $\kappa(A \cup B) = \min(\kappa(A), \kappa(B))$.

According to a NCF κ, propositions are believed in various degrees. It is useful to explicitly introduce the function expressing these degrees, because it is more vivid than the above disblief talk:[18]

DEFINITION 2. β is *the belief function associated with* the NCF κ iff β is a function from the power set of Ω into the set of integers extended by $+\infty$ and $-\infty$ such that $\beta(A) = \kappa(\overline{A}) - \kappa(A)$.[19] β is a *belief function* iff it is associated with some NCF.

Thus, $\beta(\overline{A}) = -\beta(A)$, and A is believed true or false or neither according to β (or κ) depending on whether $\beta(A) > 0$ or < 0 or $= 0$.[20]

So far, the various degrees of belief did not really play a theoretical role. But they are crucial for an account of belief revision, the central notion of which is this:

DEFINITION 3. Let κ be a NCF and A a non-empty proposition. Then *the A-part of κ* is the function $\kappa(\cdot|A)$ defined on A by $\kappa(\omega|A) = \kappa(\omega) - \kappa(A)$ for each $\omega \in A$. Again, this function is extended to all propositions by setting $\kappa(B|A) = \min\{\kappa(\omega|A)|\omega \in A \cap B\} = \kappa(A \cap B) - \kappa(A)$ for each $B \subseteq \Omega$. Finally, if β is the belief function associated with κ, we define, as in Definition 2, $\beta(B|A) = \kappa(\overline{B}|A) - \kappa(B|A)$.

Definition 3 immediately implies *the law of conjunction* that $\kappa(A \cap B) = \kappa(A) + \kappa(B|A)$ for all propositions A and B with $A \neq \emptyset$, and *the law of disjunctive conditions* that $\kappa(C|A \cup B)$ is between $\kappa(C|A)$ and $\kappa(C|B)$.[21]

The A-part $\kappa(\cdot|A)$ of κ can be viewed as a NCF with respect to the restricted possibility space A and thus as a grading of disbelief *conditional on A*. Accordingly, $\beta(\cdot|A)$ expresses degrees of belief conditional on A.

It is obvious that a NCF κ is uniquely determined by its A-part $\kappa(\cdot|A)$, its $\overline{A}$-part $\kappa(\cdot|\overline{A})$, and the degree $\beta(A)$ of belief in A. This suggests a simple model of belief revision for NCFs. If a piece of information consists only in the proposition A, then it is plausible to assume that only the old degree $\beta(A)$ of belief in A gets changed to some new degree $\beta'(A) = n$, whereas the A-part and the $\bar{A}$-part of the old NCF κ are left unchanged; n, $\kappa(\cdot|A)$, and $\kappa(\cdot|\bar{A})$ then determine a new NCF κ', the revision of the old κ by that information.[22] There are also more complicated models in which the information need not concern a single proposition. These suggestions indicate that NCFs indeed allow for a theory of revision and induction for plain belief. But there is no need to further develop the theory of NCFs.[23] The sequel requires mainly an intuitive grasp of the notions introduced in Definition 1—3.

A first useful application of these notions is the concept of a reason. Being a reason is always relative to an epistemic background, and given such a background a reason strengthens the belief in, or, in other words, is positively relevant to, what it is a reason for. This intuition can be immediately translated into formal terms:

DEFINITION 4. Let β be the belief function associated with the NCF κ, and A, B, and C three propositions. Then A is *a reason for B relative to β* (or κ) iff $\beta(B|A) > \beta(B|\bar{A})$. And A is *a reason for B conditional on C relative to β* (or κ) iff $\beta(B|A \cap C) > \beta(B|\bar{A} \cap C)$.

Note that, according to this definition, the relation of being a reason is *symmetric, but not transitive*, in analogy to probabilistic positive relevance, but in sharp contrast to the narrower relation of being a deductive reason (which is just set inclusion between contingent propositions[24]). Note also that, according to this definition, being a reason does not presuppose that the reason is actually given, i.e. believed; on the contrary, whether A is a reason for B relative to β is independent of the degree $\beta(A)$ of belief in A.

Since the value 0 has the special role of a dividing line between belief and disbelief, different kinds of reasons can be distinguished:

DEFINITION 5.

$$A \text{ is a } \left\{\begin{array}{l} \textit{additional} \\ \textit{sufficient} \\ \textit{necessary} \\ \textit{weak} \end{array}\right\} \textit{reason for } B \textit{ relative to } \beta \text{ (or } \kappa\text{) iff } \left\{\begin{array}{l} \beta(B\;A) > \beta(B\;\bar{A}) > 0 \\ \beta(B\;A) > 0 \geqslant \beta(B\;\bar{A}) \\ \beta(B\;A) \geqslant 0 > \beta(B\;\bar{A}) \\ 0 > \beta(B\;A) > \beta(B\;\bar{A}) \end{array}\right\}.$$

Conditional reasons of the various kinds are defined similarly. If A is a reason for B, it belongs at least to one of these four kinds; and there is just one way of belonging to several of these kinds, namely by being a necessary and sufficient reason. Though the emphasis will be on sufficient and on necessary reasons, the two other kinds, which do not show up in plain belief and are therefore usually neglected, well deserve to be allowed for by Definition 5.

3. CAUSATION AND EXPLANATION

Ultimately, this section will arrive at a (partial) explication of causal explanation. But this will be only a small step beyond saying how causation is related to the general account of induction for plain belief just formally introduced. So, let me turn to the latter task.

A is a cause of B, as a first approximation, iff A and B both obtain, A precedes B, and under the obtaining circumstances A raises the epistemic or metaphysical rank of B. Most people can agree on this vague characterization, the disagreement is only about how to precisely understand it. It's thus a good start; four points call for comment.

(1) 'A and B obtain': The precise nature of the causal relata A and B is a serious problem beyond the scope of this paper. I just take them to be propositions; since I did not say much about what propositions are except that they are subsets of Ω, this can hardly be wrong. No one doubts that the causal relata have to obtain, to be facts, or to be actual. This entails that causation is world-relative, i.e. that the explicandum rather is 'A is a cause of B in ω'. In the given framework, the condition that A and B obtain in ω is simply expressed by the clause that $\omega \in A \cap B$.

(2) 'A precedes B': Some think that backwards causation should not be excluded by definition, and some more think that at least instantaneous causation should be allowed. I am not sure. But since this is not my present concern, I will just stick to the temporal precedence of the cause.

But so far, there is no time in possible worlds; they need a bit more structure: Let I be a non-empty set of factors or variables; each variable $i \in I$ is associated with a set Ω_i containing at least two members; Ω_i is the set of values i may take. The set Ω of possible worlds is then represented as the Cartesian product of all the $\Omega_i (i \in I)$. Thus, each ω

$\in \Omega$ is a course of events, a function assigning to each variable $i \in I$ the value $\omega(i)$ which i takes in the possible world ω.

I shall call I a *frame* and say that Ω and its elements are *generated* by the frame I and that a NCF on Ω and its associated belief function are *for* the frame I. Already here it is clear, and to be emphasized, that the explication of causation given below will be frame-relative. This is unavoidable, if the explication is to be expressed in formally well-defined terms. Though this frame-relativity seems to me to be natural, one may find it awkward that A is a cause of B within one frame, but not within another. From the present position this relativity can only be overcome by moving into a fictitious universal frame I^* which is not further extensible. Since we shall have occasion in the next section to indulge in that fiction, we may at present be content with this relativity.

Time may now simply be represented by a weak order $\leqslant$, i.e. a transitive and connected relation, on the frame I (since metric properties of time are irrelevant); $<$ denotes the corresponding irreflexive order on I; and for $j \in I$, $I_{<j}$ is to be the set $\{k \in I \mid k < j\}$. I shall neglect the complications of continuous time and assume that time is discrete.

Time should be associated not only with variables, but, if possible, also with propositions. Therefore we define a proposition A to be a *J-measurable* or, in short, a *J-proposition* for a set $J \subseteq I$ of variables iff for all $v, \omega \in \Omega$ agreeing on J, i.e. with $v(i) = \omega(i)$ for all $i \in J$, $v \in A$ iff $\omega \in A$. Maximally specific J-propositions will be called J-states; thus, A is a *J-state* iff A is J-proposition and any two $v, \omega \in A$ agree on J. In particular, ${}^{\omega}J$ is to denote the J-state $\{v \in \Omega \mid v$ agrees with ω on $J\}$.

There are many contingent propositions which are about a single variable, and the temporal order of variables is easily carried over to them. Indeed, I see no loss at all in restricting causes and effects to be such, so to speak, logically simple propositions which are about one variable. The condition that A precedes B will thus be expressed by requiring that there are $i, j \in I$ with $i < j$ such that A is a contingent i-proposition and B a contingent j-proposition.

(3) 'A raises the epistemic or metaphysical rank of B': The clumsiness and obscurity of this phrase is due to its intended generality. That A raises the rank of B simply means that the rank of B given A is higher than given $\bar{A}$; thus the phrase makes sense only if conditional ranks are defined. With respect to probabilistic causation, these ranks

are probabilities, of course; and they are epistemic or metaphysical ranks according to whether probabilities are interpreted subjectively as degrees of belief or objectively as chances. With respect to deterministic causation, the phrase covers all kinds of approaches – regularity theories, counterfactual theories, analyses in terms of necessary and/or sufficient conditions (however these are understood in turn), etc. – which differ on the interpretation of metaphysical and epistemic ranks. What I shall propose is easily anticipated: I shall take ranks to be epistemic ranks as given by belief functions in the sense of Definition 2. Thus, this is the point where I, following Hume, trace the essential connection between induction and causation.

Why should one follow Hume and conceive causation as an idea of reflexion, as he calls it?[25] Why construe the apparently realistic notion of causation as essentially epistemically relativized? Why try to say not what causation *is*, but rather what the causal conception of a subject in a given epistemic state is? After all, Hume himself was not so unambiguous; his definition of causation as a philosophical relation is pure regularity theory void of any epistemic elements, and when stealing the realist's thunder in (1739), pp. 167–169, his emphasis is on that definition. I cannot do justice here to this profound problem, which even provoked Kant's so-called Copernican revolution; let me just mention my two main motives for taking Hume's side.

One reason is quite concrete. The literature is full of examples presenting problems to various explications of causation, and an explication of causation relative to belief functions is, I believe, more successful in coping with these problems than rival accounts. I will expand a bit on this claim after the formal explication.

The other reason is that there is not only a strong realistic intuition of causation, but also an urgent need for explaining the most prominent epistemological role of the notion of causation. If causation is epistemically relativized, this explanation ensues naturally. But without such a relativization, I do not know of a good explanation. If causation is conceived as a kind of physical ingredient of the world (say, energy transfer), the explanation would have to go like this: "There are a lot of people around, and I can't fail to notice them; therefore, people play an important role in my world picture. Similarly, there is a lot of causation around, and I can't fail to notice it; this explains the prominent epistemological role of the notion of causation." But that parallel sounds wrong to me; it underrates the peculiar epistemological importance of causation, which is different from that of people and other ubiquitous things. And if causation is conceived as a kind of structural component of the world (say, a deductive relation between laws of nature and singular facts, or a relation of counterfactuality, or a certain relation between universals), the explanation must be given in terms which cannot be accepted without further elucidation. Such terms may be lawlikeness for a regularity theory, similarity between possible words for the counterfactual account of Lewis (1973),[26] a theoretical relation of causal necessitation between universals for Tooley (1987), sect. 8.3.2, etc.; and I am not convinced that there are unproblematic ways of objectively understanding these terms.

Of course, the realistic intuition of causation should not be forgotten because of the epistemological concern. Hume did not forget it, as his two definitions of causation as a philosophical and a natural relation show, in which regularity is the objective counterpart to subjective association. Any more adequate implementation of Hume's general strategy has to make the same kind of move. In particular, it is incumbent on me to say under which conditions there is a kind of objective counterpart to NCFs or belief functions.[27] However, here I will be content with the primary, epistemically relativized explication.[28] These remarks may also make the above-mentioned frame-relativity more plausible.

(4) 'Under the obtaining circumstances': This phrase is also beset with difficulties. In particular, it seems that the relevant circumstances of A's causing B are all the other causes[29] of B which are not caused by A; and this renders the initial characterization of causation patently circular.[30] However, the circularity dissolves, if only A's being a *direct* cause of B is considered; there are then no intermediate causes, i.e. no causes of B which are caused by A, and thus the relevant circumstances may be conceived as consisting of all other causes of B. Moreover, it seems to do no harm when all the irrevelant circumstances are added, i.e. all the other facts preceding, but not causing B; and thereby the obtaining circumstances of A's directly causing B may be conceived as consisting of all the facts preceding B and differing from A. This is what I propose:

DEFINITION 6. Let $\omega \in \Omega$, $i, j \in I$, A be an i-proposition, and B a j-proposition. Then A is *a direct cause of B in ω relative to* the belief function β (or the associated NCF κ) iff $\omega \in A \cap B$, $i < j$,[31] and

$\beta(B \mid A \cap {}^{\omega}(I_{<j}\text{-}\{i\}) > \beta(B \mid \overline{A} \cap {}^{\omega}(I_{<j}\text{-}\{i\})$, i.e. A is a reason for B conditional on ${}^{\omega}(I_{<j}\text{-}\{i\})$ relative to β. More specifically, A is *an additional, sufficient, necessary, or weak direct cause of B in ω* according to whether A is an additional, sufficient, necessary, or weak reason for B conditional on ${}^{\omega}(I_{<j}\text{-}\{i\})$.

In my (1980), pp. 79ff., and (1983), pp. 384ff., I have more fully argued that ${}^{\omega}(I_{<j}\text{-}\{i\})$, i.e. the state in ω of all the variables preceding the effect and differing from the cause is indeed the correct proposition to conditionalize on; that is, I have argued that whenever we base our judgment about the direct causal relation between A and B on fewer facts, it could be just the neglected facts which would change the judgment. This is confirmed by the more detailed investigation into the relevant circumstances of causal relations in sect. 4 of my (1990a).

I believe that causation in general should be defined as the transitive closure of direct causation, as seems quite natural and as many have assumed. A fully defense of this view, however, is a long story, parts of which I have told in my (1990a). For the present purpose, it suffices to consider only direct causation.

To make Definition 6 a bit more vivid, it may be helpful to briefly explain how it deals with three standard difficulties. The first is the problem of irrelevant law specialization introduced by Salmon (1970), pp. 177ff., which says that, according to the original Hempel-Oppenheim account, John's regularly taking birth control pills explains his not becoming pregnant. Regularity theories of causation are of course threatened by this problem, too. But there is no problem for Definition 6. Given John is a man (before the given time of his non-pregnancy), his regularly taking contraceptives (before that time) is just irrelevant to, and not a reason for, his non-pregnancy at that time, at least relative to our educated belief functions.

The second problem is the distinction between causes and symptoms which is a graver obstacle to regularity accounts of causation and explanation. Take, e.g., C. D. Broad's Manchester hooters and London workers discussed at length by Mackie (1974), pp. 81ff. Whenever the factory hooters in Manchester and London sound, which is the case every working day at 6 p.m., then the workers in Manchester and London leave their work shortly afterwards. But only the London and not the Manchester hooters have an impact on the London workers. Again, this case presents no problem for Definition 6. Unconditionally, the proposition that the Manchester hooters sound (at a particular day) is relevant to the proposition that the London workers leave; but given that the London hooters sound (or do not sound), the former is just irrelevant to the latter. Again this is true relative to our normal belief functions. Of course, one may have a different belief function yielding also a conditional relevance; but then the sounding of the Manchester hooters is not treated as a mere symptom of the London workers' leaving.

The general scheme should be clear by now. NCFs and belief functions help us to notions of (conditional) relevance and irrelevance which are much more sensitive than the relevance notions provided by other approaches to deterministic causation and which behave almost the same as probabilistic relevance notions.[32] Thus, they enable us to copy the methods of dealing with these problems which have been so successfully developed for probabilistic causation.

A third problem further illustrating this scheme distinguishes Definition 6 not only from regularity theories, but also from counterfactual analyses of causation like Lewis (1973). It is the problem of (symmetric) causal overdetermination cruelly, but standardly, exemplified by the firing squad. Prima facie, cases of causal overdetermination are clearly possible. But as far as I see, they are a great mystery, if not an impossibility for all realistic accounts of causation; it seems that the only thing the realist can do is to explain them away: either by observing that one of the two causal chains from the two apparently overdetermining causes to the effect has not been completed so that one of the two causes is in fact preempted; or by observing that there is an intermediate event (a Bunzl event, as Lewis (1986), p. 208, calls it) which causes the effect and which is jointly caused by the two apparently overdetermining causes so that the two causes in fact jointly cause the effect.[33]

For Definition 6, however, there is no mystery at all. The two overdetermining causes may be simply conceived as additional causes; each of the two is an additional cause of the effect in the presence of the other one. The crucial difference is that additional causes cannot be defined within a counterfactual approach, let alone a regularity theory. Something true can counterfactually be still true or not true, but not more or less true. But something conditionally believed can be believed more or less firmly under different conditions.[34] Of course, I do not claim that this simple remark solves the problem of causal overdetermination. What it does is first to do justice to the prima facie existence of causal overdetermination and secondly to locate the problem; it arises when we try to objectivize our epistemically relativized causal picture, because there is no realistic counterpart to additional causes as defined above.

So much for the partial account of causation we need. It is easily extended to a partial account of explanation. I shall not comment on explanation of laws and theories and on whether there is non-causal explanation.[35] But concerning causal explanation, it seems unassailable to say that getting an explanation for B is learning a cause of B and having an explanation for B is knowing a cause of B.[36] The problem is only that this statement is unhelpful as long as one does not have an account of causation or tries to explain causation by explaining explanation. But this is not our problem, and thus we may immediately turn this informal statement into a formal definition:

Knowing some fact to be a cause at least involves believing this fact to be a cause. And believing A to be a cause of B according to a NCF κ means believing the actual world to be among the worlds in which A is a cause of B relative to κ. Since only direct causes have been formally defined, this leads to

DEFINITION 7. Let i, j, A, and B be as in the previous definition. *A's range* $C_{A,B}$ *of directly causing B relative to* the NCF κ or its associated belief function β is defined as the $I_{<j}$-$\{i\}$-measurable set of all $\omega \in \Omega$ such that $\beta(B \mid A \cap {}^{\omega}(I_{<j}\text{-}\{i\}) > \beta(B \mid \bar{A} \cap {}^{\omega}(I_{<j}\text{-}\{i\})$. Hence $A \cap B \cap C_{A,B}$ is the set of all $\omega \in \Omega$ such that A is a direct cause of B in ω relative to κ or β.[37] *A's range* ${}^{s}C_{A,B}$ or ${}^{n}C_{A,B}$ *of*, respectively, *sufficiently or necessarily directly causing B relative to* κ or β is defined accordingly. Then, *A causally explains B* (*as necessary, as possible*) *relative to* κ or β iff $\beta(A \cap B \cap C_{A,B}) > 0$ $(\beta(A \cap B \cap {}^{s}C_{A,B}) > 0$, $\beta(A \cap B \cap {}^{n}C_{A,B}) > 0)$.[38]

The only deviation of Definition 7 from its informal statement is that knowledge of a cause has been weakened to belief in a cause. This corresponds to the old debate whether explanation requires true or only accepted antecedents. I think there are both usages; 'B is explained by A' may be factive or not according to whether it is taken as the passive of the apparently factive 'A explains B' or as an ellipsis of the apparently non-factive 'B is explained by A by some explainer'. Since I have always talked only of belief and not of knowledge, I settle for the weaker version. I do not see that our topic is seriously affected by this issue. In particular, I do not see that this issue drives a wedge between explanation and understanding, as Lambert (1988), pp. 308–310, and (1991), pp. 138f., argues. Understanding as well can be taken factively or non-factively, and it seems only fair that, when assessing the relation between explanation and understanding, only the corresponding interpretations are compared.

4. REASON AND TRUTH

The first task of giving a partial account of explanation need not be developed further.[39] The next task is to give an independent account of understanding or, rather, of some not too bad substitute thereof. I approach this task by discussing three principles of increasing strength which I take to be basic principles of coherence, believability, and truth.

Let's start with a simple question: If B is a contingent proposition, is there a reason for B? Trivially, yes. There always are deductive reasons; each non-empty subset of B is a sufficient, and each non-tautological superset a necessary, deductive reason for B. So, the question should rather be whether there is an *inductive*, i.e. non-deductive reason for B. Or, put in another way, if B is a contingent j-proposition for some $j \in I$, is there a I-$\{j\}$-proposition which is a reason and thus an inductive reason for or, for that matter, against B?[40] Not necessarily, of course. There may be variables which are independent of all other variables in the given frame I relative to the given belief function β; and since I may be an arbitrary collection of variables, such counterinstances ensue naturally.

Consider now an extension I' of the frame I and an extension β' of β for I'. There are many such extensions of β, and, trivially, there exists an extension of β according to which there is a I'-$\{j\}$-measurable reason for B. Thus, we should, more precisely, consider *the* extension β' of β as determined by some unspecified epistemic subject with a belief function covering also I'-propositions. Is there a I'-$\{j\}$-measurable reason for B according to β'? Again, not necessarily. The case of I' is not different from the case of I.

But now consider all extensions I' of I and the appertaining belief functions β' which are within the subject's range. Should then an inductive reason for or against B come to the fore? Once more, not necessarily; but that would be a grave matter. It would mean that the subject could not learn anything whatsoever about B; wherever it looked, it could not find the slightest hint concerning B; B would be outside its world of experience, outside its bounds of sense.

It may of course happen that a proposition is beyond a subject's present grasp. This may change; an individual's inductive scheme as well as that of a society or of mankind keeps evolving.[41] It may even be that a proposition is forever beyond the grasp of an individual or of actual mankind. But these are all accidental limitations. My real concern is the status of the j-proposition B with respect to all possible extensions of the frame I whatsoever and the appertaining belief functions to which a subject would extend its belief function β, if it came to consider these extensions of I. Is it still conceivable that in this sense no extension of I contains an inductive reason for B? Now, finally, it seems plausible to say no. Otherwise, there would be no way whatsoever to reason or to learn anything about B, not because of accidental limitations, but due to the inherent structure of the all-inclusive inductive scheme underlying all these extensions of β; B would be literally senseless, unreasonable.

I have referred to all possible extensions of some initial frame and inductive scheme. But it is simpler to refer instead to the universal frame I^* comprising all variables whatsoever, to the set Ω^* of possible worlds generated by I^*, and to a universal belief function β^* for I^*. It may seem earthlier to talk only of extensions. But the set of all extensions is not earthlier than its union; both are philosophical fiction. Talking of I^*, Ω^*, and β^* is just much less clumsy than quantifying over extensions.

I^*, Ω^*, and β^* are what, in a loosened usage of Kantian and Peircean terms, has been called regulative ideas, ideal limits of inquiry, etc. The question whether one can legitimately and sensibly appeal to such limit concepts is certainly pressing. Here I just follow all those who do so. And I take it that, insofar our epistemic activities may at all be described by frames and belief functions, we conceive these activities as embeddable into the universal frame I^* and a universal belief function β^* and that we consider this embeddability as a fundamental requirement of consistency.[42]

What we have arrived at, then, is a first plausible principle of coherence:

(PCo1) For any $j \in I^*$ and any contingent j-proposition B there is a I^*-$\{j\}$-measurable reason for B relative to β^*.

Pco1 may be taken as a condition on β^*, on how β^* has to connect propositions. But it may also be conceived as a condition on I^* (and the generated Ω^*) saying that no logically simple proposition exists unless appropriately connected by β^*. The best is to view PCo1 as what it is, as a condition simultaneously on I^* *and* β^*.

PCo1 is, of course, akin to the positivists' verifiability principle and other criteria of empirical significance. But PCo1 is a weak version, because it requires at best confirmability and not verifiability and because it does not refer to a directly verifiable basis, to evidential certainty, and the like. And PCo1 is unambiguous about the nature of the required ability of confirmation. This ability is not to be taken as restricted by our sensory outfit; PCo1 does not refer to any specific senses. It is not restricted by limited computing capacities; β^* will not be computationally manageable, anyway. It is not restricted by our spatiotemporally and causally limited access to facts. This ability is constituted exclusively by the inherent structure of the limiting inductive scheme and thus of the actual inductive schemes approaching it.

Given the above explication of direct causation, PCo1 is, by the way, tantamount to the following weak principle of causality:

(PCa1) For any $j \in I^*$ and any contingent j-proposition B there is a direct cause or a direct effect of B in some world $\omega \in \Omega^*$ relative to β^*.

At least the equivalence of PCo1 and PCa1 holds, if I^* is linearly and discretely ordered by time.[43] Note also that the reference to I^* and β^* eliminates the frame-relativity of that explication, but not its epistemological involvement.

PCo1 is symmetric with respect to positive and negative relevance; whenever a proposition is a reason for B, its negation is a reason against B. This symmetry will break in the next step when we consider *true* propositions; truth is biased towards positive relevance:

We have first to introduce another limit concept: the actual world taken not as a spatiotemporally maximally inclusive thing, but as everything that is the case. We naturally assume that among all the possible worlds in Ω^* exactly one is actual; let's call it α^*. Thus, a proposition A is true (absolutely, not relative to a model or a world) iff A is true in α^*, i.e. iff $\alpha^* \in A$.

The question now is this: Suppose that the contingent j-proposition B is true. PCo1 asserts that there are I^*-$\{j\}$-measurable reasons for B. But will there be a true reason among them? Let's look at the question in a more earthly setting of a small frame I, the small actual world α (which is the restriction of α^* to I), and a subject's belief function β for I. Within this setting, the answer may certainly be no. If so, however, the truth would be undetectable, unbelievable for the subject within this setting. If it believed only truths, it would have no reason for believing B; and if it has any reason for believing B, then only by believing some I-$\{j\}$-propositions which are false. This situation is not critical by itself, but it again becomes more and more critical when it does not change as larger and larger extensions of I are considered. And relative to I^*, α^*, and β^*, finally, this situation seems absurd; all true evidence which could conceivably be brought to bear on B would univocally speak against B and for $\overline{B}$, though B is true and $\overline{B}$ false. Thus it seems plausible to answer the initial question in the affirmative.

This can be stated as a second principle of coherence:

(PCo2) For any $j \in I^*$ and any contingent j-proposition B with $\alpha^* \in B$ there is a I^*-$\{j\}$-measurable proposition A with $\alpha^* \in A$ which is a reason for B relative to β^*.

Briefly: for each singular truth there is a true inductive reason. Of course, PCo2 implies PCo1.

In Peirce-Putnamian terms one might say that PCo2 is part of the assertion that the epistemically ideal theory cannot be false. The ideal theory has, of course, recourse to all true evidence; and in a case violating PCo2 the ideal theory would have to falsely affirm $\overline{B}$ on the basis of that evidence and the universal inductive scheme β^*. PCo2 prevents this and thus captures at least one aspect of Putnam's internal realism.[44]

Indeed, PCo2 fits well under the heading 'coherence theory of truth'. The theoretical standing of the coherence theory is not exactly brilliant, because of difficulties in saying precisely what coherence is. Explications in deductive terms, say as consistency or deducibility, were precise, but unprofitable; and other, more interesting explications were always vague. A noticeable exception is Rescher (1973) and (1985); but I find his underlying theory of plausibility indexing not fully satisfying. Here, coherence is construed as inductive coherence as constituted by positive relevance relative to a belief function. PCo2 is thus one way of saying that truth must cohere. Of course, a workable theory of induction or belief revision for plain belief is vital to this construal.

PCo2 does certainly not yield a definition of truth. For propositions, being true is defined as having α^* as a member; and for sentences, Tarski's truth definition may need an underpinning by a theory of reference, as called for by Field (1972), but as a definition it does not need a coherentist supplement. PCo2 also does not yield a criterion of truth; it is of little help in determining the truth of B because it is kind of circular in requiring true reasons for B and because it does not tell what to do in the case of conflicting reasons. In fact, PCo2 is not a condition on truth alone; it must again be viewed as a condition on I^*, α^*, *and* β^*, on how truth and reason relate in the universal frame.

There is also a principle of causality associated with PCo2:

(PCa2) For any $j \in I^*$ and any contingent j-proposition B with $\alpha^* \in B$ there is a direct cause or a direct effect of B in α^* relative to β^*.

Briefly: each singular fact has a direct cause or a direct effect in the actual world. This principle of causality is, of course, much stronger and much more interesting than PCa1. PCa2 is even stronger than PCo2; the former implies the latter, but not vice versa.[45] It would be nice to find a plausible principle of coherence entailing PCa2; so far I have not succeeded.

There are, however, plausible strengthenings of PCo2. One of them is my next goal.

PCo2 asserts that a true I^*-$\{j\}$-measureable reason A may be found for the contingent true j-proposition B. Now imagine that a piece C of true information is received and that A is then no longer a reason for B, i.e. A is not a reason for B conditional on C. This is not impossible; if A is positively relevant to B given one condition, A may be positively, negatively, or not relevant to B given another condition. And it is not excluded by PCo2. But this seems an implausible way to satisfy the plausible PCo2.

This opens up a new kind of question: How does the relevance of some truth to B evolve in the infinite process of acquiring more and more true information? Formally, everything is possible. The relevance may (a) vacillate for some (or no) time and then forever stay on the positive side, or (b) vacillate for some (or no) time and then forever stay on the non-positive side, or (c) vacillate forever. A truth of kind (b) is a very casual kind of reason for B, if at all, and one of kind (c) an odd and deeply undecided kind.

Is it conceivable that all true reasons for B one finds after some true information or other turn out to be of these unreliable kinds (b) and (c)? Formally, there are again three ways how this might happen. First, it might be that true reasons for B run out after sufficient true information. This case definitely violates the basic idea of PCo2 that in the limit all truth must be believable. Secondly, it might be that at infinitely many stages of the acquisition of true information there are true reasons for B and at infinitely many other stages there are no true reasons for B. This case again violates the basic idea. As often as one gains confidence in B, one loses it; one can never hold it fast. Thirdly, it might be that after some true information there always are true reasons for B, though different ones at each subsequent stage of the process. This case seems to be compatible with the basic idea, but it is still strange. Each time when asked why one believes B one has to withdraw the previous answer and to give another one; and this continues forever. This does not seem to be an acceptable process of truth tracking.

I therefore conclude that there should be at least one reason for B of the reliable kind (a); I shall call it an *ultimately stable* reason. This is the key concept of the following considerations; it is more precisely defined thus:

DEFINITION 8. Let $\omega \in \Omega$ and $A, B, C \subseteq \Omega$. Then A is a *ω-stable* (*sufficient, necessary*) *reason for B within C relative to* a belief function β for I (or the associated NCF κ) iff $\omega \in A \cap B$, $\omega \in C$, $A \cap C \neq \emptyset \neq \bar{A} \cap C$, and A is a (sufficient, necessary) reason for B relative to β conditional on each $D \subseteq C$ with $\omega \in D$ and $A \cap D \neq \emptyset \neq \bar{A} \cap D$. A is an *ultimately ω-stable* (*sufficient, necessary*) *reason for B relative to β* iff A is so within some condition. The set of all $\omega \in \Omega$ such that A is an ultimately ω-stable (sufficient, necessary) reason for B is called *A's range of being an ultimately stable* (*sufficient, necessary*) *reason for B* and denoted by $S_{A,B}$(${}^{s}S_{A,B}$, ${}^{n}S_{A,B}$).

Note that the truth of A and B in ω is made a defining characteristic of A's being an ultimately stable reason for B. Note also that, if A is an ultimately ω-stable reason for B, so is B for A.

In these terms, then, I have just argued for a third principle of coherence:

(PCo3) For any $j \in I^*$ and any contingent j-proposition B with $\alpha^* \in B$ there is a I^*-$\{j\}$-measurable, ultimately α^*-stable reason for B relative to β^*.

Briefly: for each singular truth there is an ultimately stable inductive reason. If there are reasons with stronger than ultimate α^*-stability, say, with α^*-stability within Ω, all the better. But such stronger forms of stability do not seem to be required in PCo3 on coherentistic grounds. Still, PCo3 implies PCo2.[46]

5. EXPLANATIONS AND STABLE REASONS

Now I can finally offer my substitute for (scientific) understanding: it is knowing ultimately stable reasons. I do not want to defend this as an explication of the complex notion of understanding. But what has been said in this context about understanding is captured fairly well by my proposal; and knowing ultimately stable reasons is epistemologically significant in its own right. Let me explain.

What is meant by knowing an ultimately stable reason A for B? Not only that one knows A and A is in fact an ultimately stable reason for B, but also that one knows A to be so, i.e. that one knows A's range $S_{A,B}$ of being an ultimately stable reason for B to obtain. As in the case of explanation, there is a factive and a non-factive understanding of understanding, and as in the former case I deal only with the latter, in order to be able to confine myself to belief and to be silent about knowledge. To be precise, then, A's being believed to be an ultimately stable reason for B relative to β simply comes to $\beta(S_{A,B}) > 0$.

The significance of believing in ultimately stable reasons is this: When one believes A and B to be true, one thinks that A and B are part of, fit into, α^* in some way or other. But one may do so as a mere recorder of facts without any understanding of what is going on, without any grasp of *how* A and B fit into α^*. And one may, adhering to PCo3, simply proclaim that it should be possible to find an ultimately α^*-stable reason for B. When one believes $S_{A,B}$ to be true, however, one does not only believe A and B, and one does not merely postulate an ultimately α^*-stable reason for B. Rather, one thinks to know a particular one, namely A. And one has a partial grasp of how A and B fit into α^*, namely as one element of coherence, as one coherent link among many others which have to exist in α^*. Thus, for the believer of $S_{A,B}$ A and B better qualify as part of the final truth than for the believer of A and B alone.

How else is understanding characterized? Lambert (1991), p. 129, says that “the metaphor of ‘fitting into’, and its stylisic variants such as 'incorporated into' or 'integrated into', seem especially germane vis à vis scientific understanding as it relates to scientific explanation" and quotes a number of important authors using this metaphor. For him, then, "the sense of scientific understanding relevant to scientific explanation may be characterized as an answer to the question 'How does state-of-affairs *S* fit into theory *T*?'" (p. 130), where, as he goes on to explain, "fit into" may mean various things.

Similarly, Friedman (1974) and Kitcher (1981), again adducing a number of witnesses, take unification as the key concept. Friedman (1974), p. 15, explicitly claims:

> ... this is the essence of scientific explanation — science increases our understanding of the world by reducing the total number of independent phenomena that we have to accept as ultimate or given. A world with fewer independent phenomena is, other things being equal, more comprehensible than one with more.

And he goes on to say more precisely how he understands independence or independent acceptability.

These seem to be appropriate ways of talking also about ultimately stable reasons; indeed, I myself slipped into these ways three paragraphs ago. Of course, the metaphors take on different senses with the different authors. But this is not an unhappy homonymy which hides incomparable interests. On the contrary, I think, there is one common idea which is vague and allows of various explications, and there is less a disagreement about its content, but rather a common need in surveying this idea and tracing fruitful explications. Here, in any case, fit and unification, like coherence, are construed as inductive fit and unification as constituted by (conditional) positive relevance relative to a belief function.

On a strategic level, the main difference between the papers referred to and the present proposal is that there fit and unification are somehow construed as relations between facts or phenomena and theories, whereas here they are construed as a relation between facts and inductive schemes. Talking of theories is certainly closer to scientific practice, but talking of inductive schemes is nearer to epistemological theory. Is there a substantial difference? This is unclear as long as the relation between theories and inductive schemes is not made clear. Without doubt there is a close relation, and it is incumbent on me to say how theories are implicitly contained in inductive schemes; I shall not attempt to do so here. Conversely, however, there is an urgent need to say how inductive schemes are implicitly contained in scientific theories; I am convinced that mere reference to theories is not helpful for all the topics here addressed as long as theories are conceived as something modally inert, e.g. as sets of extensional sentences or extensional models.[47]

These remarks also suggest an answer to the question on which Salmon (1978) hangs his discussion. Salmon asks on p. 684:

> Suppose you had achieved the epistemic status of Laplace's demon ... who knows all of nature's regularities, and the precise state of the universe ... at some particular moment. ... you would be able to predict any future occurrence, and you would be able to retrodict any past event. Given this sort of apparent omniscience, would your scientific knowledge be complete ...? Laplace asked no more of his demon; should we place further demands upon ourselves?

In the sequel Salmon explains what Laplace's demon lacks. From the present point of view, omniscience — whether it is direct as presumably that of God or inferred from a complete set of axioms as that of Laplace's demon — is neither an ideal nor a counterfactual epistemological possibility for us. The reason is not that it is impossible on various scores to know so much. The reason is rather that we could not merely know everything; having an inductive scheme, proceeding inductively in the broad sense here always referred to is an essential and indispensible feature of our epistemic constitution which would not fade by approaching omniscience. Laplace's demon is indeed granted too little; it would not know what to believe, if it were to discover that it is wrong. We would know, even while approaching omniscience. If I am right, all the other things which the demon is held to be wanting in this discussion including those mentioned by Salmon (1978), p. 701, result from this central lack.[48]

Having thus shed some light on the epistemological locus of stable reasons, I can finally turn to the object of my paper: the relation between explanations and ultimately stable reasons. Though the definitions of $C_{A,B}$ (Def. 7) and of $S_{A,B}$ (Def. 8) look quite similar, this relation is not straightforward. The main difference is that direct causes are characterized by conditionalization on the whole past of the effect, whereas ultimately stable reasons are characterized by conditionalization on many and finally all other truths, whether past or future. This prevents a direct comparison. There is help, however: just restrict all the coherentistic considerations about the j-proposition B in the previous section to the past of B. This move brings easy success, indeed too easy, and therefore two disappointments. I shall first state in precisely what the move and its success consist, before explaining what the disappointments are and what might be done about them.

The move is simple: Restate PCo1 as saying that for any $j \in I^*$ and any contingent j-proposition B there is an $I^*_{<j}$-measurable reason for B relative to β^*. This is equivalent to a modified PCa1 saying that for any such j and B there is a direct cause of B in some world $\omega \in \Omega$ relative to β^*. Change PCa2 and PCo2 in the same manner; the former is again implied by the latter.[49] Modify finally Definition 8: Define A to be a *ω, j-past-stable (sufficient, necessary) reason for B within C relative to β* by additionally requiring C to be $I_{<j}$-measurable and by requiring A to be a (sufficient, necessary) reason for B relative to β conditional only on each $I_{<j}$-measurable $D \subseteq C$ with $\omega \in D$ and $A \cap D \neq \bar{A} \cap D$; define accordingly A's being an *ultimately ω, j-past-stable (sufficient, necessary) reason for B* and A's *range of being an ultimately j-past-stable (sufficient, necessary) reason for B*; and denote this range by $S_{j;A,B}({}^sS_{j;A,B}, {}^nS_{j;A,B})$. PCo3 may then be reformulated correspondingly.

After this modification the comparison is immediate: If A is an i-proposition and B a j-proposition with $i < j$ and if i is a binary variable,[50] then $A \cap B \cap C_{A,B} = S_{j;A,B}$ $(A \cap B \cap {}^sC_{A,B} = {}^sS_{j;A,B}$, $A \cap B \cap {}^nC_{A,B} = {}^nS_{j;A,B})$ and thus A causally explains B (as necessary, as possible) relative to β if and only if A is believed to be a j-past-stable (sufficient, necessary) reason for B relative to β. For proof one has only to look at the definitions and to observe first that ${}^\omega(I_{<j}\text{-}\{i\})$ is *the* smallest $I_{<j}$-proposition C with $\omega \in \Omega$ and $A \cap C \neq \emptyset \neq \bar{A} \cap C$, if i is binary, and secondly that being an ultimately ω,j-past-stable reason only requires being a reason conditional on this smallest proposition C.

Hence the justification of explanation I propose runs as follows: On the one hand, there is the explication of direct causation and consequently of causal explanation in section 3. On the other hand, there are the independent coherentistic considerations of section 4 which suggest that truth is tied to ultimately stable reasons, as stated in PCo3, and that believing in ultimately stable reasons is thus an indispensible ingredient of having a true world picture. And, as has turned out now, it is explanations and only explanations which provide these ingredients, at least if the relation of being a reason is considered only with respect to pairs of logically simple propositions about single variables and if the coherentistic considerations are restricted to the past of the later proposition of such a pair.

Somehow, however, the last step appears too trivial. It falls short of the expectations I have created in two respects:

One disappointment is that in the short proof of the identity of $A \cap B \cap C_{A,B}$ and $S_{j;A;B}$ being an ultimately stable reason takes on an unexpectedly weak sense. According to my definition, being an ultimately ω(j-past)-stable reason boils down to being a reason conditional on, sloppily put, all the rest of the truth in (the j-past of) ω. But according to my motivation in the previous section, the idea rather was that an ultimately ω-stable reason is a reason after some finite information true in ω and stays a reason after all further information true in ω.

This deficiency can be cleared, however, because the cause specified in a causal explanation is in fact a reason which is stable within the cause's range and not only from some remote point onward. More precisely, the following assertion holds true: If A is an i-proposition and B a j-proposition with $i < j$, then A is a sufficient reason for B conditional on each non-empty, $I_{<j}$-$\{i\}$-measurable $D \subseteq {}^sC_{A,B}$ relative to β.[51] Hence, for each $\omega \in A \cap B \cap {}^sC_{A,B}$, A is a ω,j-past-stable sufficient reason for B not only ultimately, but within no less than ${}^sC_{A,B}$. The assertion with 'necessary' and '${}^nC_{A,B}$' replacing 'sufficient' and '${}^sC_{A,B}$' holds correspondingly.[52] However, the assertion fails to generally hold for reasons and direct causes simpliciter.[53]

Do explanations also provide unconditional reasons? Under mild assumptions yes, provided only sufficient or necessary reasons are considered. More precisely, the following assertion holds true: If A causally explains B as necessary relative to β and $\beta({}^sC_{A,B} | \bar{A}) \geqslant 0$, then A is a sufficient reason for B relative to β[54]; and if A causally explains B as possible relative to β and $\beta(B\text{-}{}^nC_{A,B} | \bar{A}) < 0$, then A is a necessary reason for B relative to β.[55] This assertion, or at least its 'sufficient'-part, very much resemble the thesis "that an adequate answer to an explanation-seeking why-question is always also a potential answer to the corresponding epistemic why-question"[56] and may thus be taken as a proof thereof. The additional premise of the 'sufficient'-part that $\beta({}^sC_{A,B} | \bar{A}) \geqslant 0$ will usually be satisfied, I think; and one might argue that it is just this premise which is violated in alleged counter-examples to that thesis.[57]

The other disappointment is the restriction of the coherentistic considerations about the j-proposition B to the past of B in the way specified above. This is disappointing because thus restricted these considerations lose much of their persuasiveness. I have great confidence in PCo1-3 as I have stated them in the previous section; but I do not know how to convincingly argue for PCo1-3 as modified in this section. The modified PCo1-3 (and in particular the principles of causality associated with them) still look desirable, but it is not clear why they should be necessary on coherentistic grounds. This is a gap in my argument.

Perhaps, however, this unsupported restriction of the coherentistic considerations is not really necessary. How is it possible that the i-proposition A is a direct cause of the j-proposition B in ω and thus an ultimately ω,j-past-stable reason for B, but not an ultimately ω-stable reason for B? The only possibility is that some true information about the future of B turns the positive relevance of A for B given the rest of the past of B into irrelevance or negative relevance. But there is something odd about this possibility. Consider a simple formal example:

Let ω, A, B, and C be such that A precedes B, B precedes C, and $\{\omega\} = A \cap B \cap C$. Now suppose on the one hand that A is a sufficient reason for B and thus also a sufficient direct cause of B in the small world ω, and on the other hand that A is a necessary reason for $\overline{B}$ given C and thus not an ultimately ω-stable reason for B. These assumptions imply: First, $\beta(C|A) < 0$; thus A and C cannot both believed to be true, and A is at best a weak reason for C. Secondly, C is a necessary reason for $\overline{B}$ given A and, because of symmetry, B a reason for $\overline{C}$ given A.[58] Hence, C very badly fits A and B under these assumptions.

Similar assumptions create similar oddities. This suggests a general conclusion which looks at least plausible: If for all $j \in I$ true j-propositions cohere with all past truths, then, for any $i \in I$, a true i-proposition coheres with all other truths, because it coheres with all past truths, as just stated, and also with true j-propositions for all $j > i$, since coherence is symmetric. In this way general coherence with the past seems to be equivalent with general coherence with past *and* future. If this is true, the above restriction of the coherentistic considerations would, after all, not really be a restriction. However, this is only a vague conjecture, neither precisely stated nor proved.

If the presented line of reasoning is correct, we ask 'why?', we search for explanations because this is one and, in a way, the only way of finding coherent truth and, insofar as truth must be believable and coherent, the only way of finding truth. Why search for truth? Here I cannot think of any further theoretical justification; to some extent we seem to be intrinsically curious beings. Papers must end, justifications presumably, too. But the present one does not end here; there is beautiful further justification for the search for truth of a practical, decision-theoretic kind.[59]

Department of Philosophy,
University of Bielefeld,
Bielefeld, Germany.

NOTES

* I am very grateful to Dan Hunter for checking my English.

[1] Sketchy remarks about the utility of explanations may be found quite often. Much less often the question is explicitly addressed, e.g. by Salmon in (1978), where he propounds his own answer to the question, and in (1984), pp. 124ff. and 259ff., where he discusses also other answers.

[2] This is the declared strategy of, e.g., Friedman (1974) and Kitcher (1981).

[3] Indeed, this paper originated from an observation of this fit.

[4] Cf., e.g., Hume (1739), pp. 94ff., and (1777), pp. 47ff. The struggle is most conspicuous in the appendix of (1739), pp. 623ff.

[5] In (1777), p. 26, Hume writes: "All reasonings concerning matter of fact seem to be founded on the relation of *Cause and Effect.* By means of this relation alone we can go beyond the evidence of our memory and senses." In (1739), p. 107, he says equally clearly: ". . . we find by experience, that belief arises only from causation, and that we draw no inference from one object to another, except they be connected by this relation". However, the relevance of further principles of association, namely resemblance and contiguity, is not really clear. In (1739), pp. 107ff., he argues that these are only assisting, but not basic principles; in (1777), pp. 50ff., he does not discriminate in this way.

[6] This is very explicit in Hume (1739), pp. 170—172.

[7] This observation raises questions: Does Hume take one of the two notions as primary? Or is there a circularity in Hume's account? Which role has Hume's definition of causation taken as what he calls a philosophical relation, which refers to regularity instead of association? Does it offer a way out of the possible circularity? Cf., e.g., Mackie (1974), ch. 1, and Beauchamp, Rosenberg (1981), ch. 1, for thorough discussions of these questions.

[8] This is the lesson, for instance, of Goodman's new riddle of induction and Carnap's acknowledged failure to distinguish even a small class of inductive methods. It is challenged, however, by the puzzling alternative set up in the final section of Lewis (1980).

[9] Cf., e.g., Hunter (1991) in this volume.

[10] The further conclusion that plain belief is an illusion is unwarranted; it is drawn only in default of a more appropriate representation of epistemic states.

[11] See Shackle (1961), Rescher (1976), and Cohen (1977).

[12] He has summarized his work in Gärdenfors (1988).

[13] The only point of this technical label is that it be not confused with other notions. Perhaps the more suggestive term 'disbelief function' would be better, as Shenoy (1991) has proposed.

[14] My account in (1988) is slightly more general insofar as the range there consists of ordinal numbers.

[15] Setting $\kappa(\emptyset) = \infty$ is a reasonable convention. But ∞ should not be allowed as value of possible worlds and consistent propositions because no good rules of belief revision can be devised for it.

[16] The various problems which cast serious doubt on the idea that belief takes propositions as objects are pressing, but must here be disregarded.

[17] $\bar{A}$ of course denotes the complement of A relative to Ω.

[18] I thereby follow a proposal of Shenoy (1991).

[19] This definition which is much simpler than my original one has been pointed out to me by Bernard Walliser. Note that because of the law of negation at least one of the terms of the definiens is 0.

[20] The reason why the more vivid belief functions are introduced only as a derivative concept is that their formal behavior is less perspicuous.

[21] This holds because $\kappa(C \mid A \cup B) = \kappa((C \cap (A \cup B) - \kappa(A \cup B) = \min[\kappa(A \cap C, \kappa(B \cap C)] - \min[\kappa(A), \kappa(B)]$ and because $\min[y_1, y_2] - \min[x_1, x_2]$ is always between $y_1 - x_1$ and $y_2 - x_2$.

[22] In my (1988), p. 117, I have defined this process as A,n-conditionalization.

[23] For further details see my (1988) or my (1990b). There it is made clear why, given certain assumptions, revision schemes for plain belief have to take the form of NCFs; it is shown that NCFs behave very much like probability measures with respect to conditionalization and (conditional) independence; and the justification for more general forms of conditionalizations of NCFs closely parallels that for Jeffrey's generalized probabilistic conditionalization given by Teller (1976).

[24] A proposition A is contingent iff $\emptyset \neq A \neq \Omega$.

[25] This is Hume's most influential conclusion of the crucial Section XIV of Book I of his (1739). "The idea of necessity arises from some impression. . . . It must . . . be derived from some internal impression, or impression of reflexion", he writes on p. 165, and it is clear that necessity here includes causal necessity.

[26] If taken subjectively, Lewis' similarity relations are similar, but not equivalent to my NCFs; see my (1988), p. 127.

[27] My (1992) is an attempt to meet this obligation and thus to do justice to the realistic intuition within the present framework.

[28] In the third paragraph of this section I claimed to have a neutral usage of 'proposition'. This may seem objectionable because I assume propositions to be objects of belief *and* of causation and thus to play a double role which is arguably unsatisfiable. This is indeed a problem. But the problem arises within a realistic conception of causation and is thus part of and additional burden to the objectivization problem just put aside.

[29] It should be clear that a fact may have several causes. Thus I follow the common understanding which construes 'cause' as 'partial cause' and not as 'total cause'.

[30] This is precisely the idea and the conclusion of Cartwright (1979) concerning probabilistic causation.
[31] Even for direct causation it would not be reasonable to generally require that i immediately precedes j, because the frame I may be any wild collection of variables; in particular, I may contain many variables which are temporally, but not causally, between i and j. This requirement at best characterizes nice frames. It should be noticed that the assumption of discrete time is important for Definition 6; given continuous time, direct causes presumably do not exist or are simultaneous with their effects.
[32] Cf. my (1988), section 6.
[33] See the enlightening discussions of Bunzl (1979) and Lewis (1986), pp. 193—212.
[34] This, by the way, is also a point of difference between Gärdenfors' belief revision model and mine. Gärdenfors has plain conditional belief, but not more or less firm conditional belief and therefore nothing like additional reasons and causes. Cf. Gärdenfors (1988), ch. 3 and 9.
[35] However, I tend to join Lewis (1986), pp. 221ff., in thinking that there is no non-causal explanation of singular facts.
[36] This is essentially also what Lewis (1986), pp. 217ff., maintains, though he points out that knowing a cause of an event is not the only way of having information about the causation of that event. I neglect here the other ways.
[37] In section 4 of my (1990a) I have called $C_{A,B}$ the actually relevant circumstances of (the direct causal relation between) A and B in the widest sense.
[38] The need to consider explanations by additional or weak causes will not arise; thus I did not formally introduce them.
[39] It would be useful to extend my comparative remarks about causation to some remarks about how Definition 7 relates to other accounts of explanation; but this is beyond the scope of this paper. Just this much: my account seems to me to fit van Fraassen's theory in sect. 5.4 of his (1980) insofar as Definition 7 tries to say more about van Fraassen's relation of explanatory relevance for the case of direct causal explanation.
[40] It would not be reasonable to ask without restrictions on measurability whether there is an inductive reason for any contingent proposition B not of the form $\{\omega\}$ or Ω-$\{\omega\}$, because the answer turns out to be yes whenever Ω has more than four members.
[41] However, I don't know of any theory about the evolution of inductive schemes, i.e. about the change of belief functions, probability measures, or whatever *for changing frames*, except of conceiving it as generated by an underlying, more inclusive inductive scheme.
[42] Cf. also the quite similar remarks of Ellis (1979), pp. 9ff., about what he calls the ideal of completability.
[43] This is easily proved on the basis of two properties of conditional independence between sets of variables which are stated as assertion (7) in my (1990b) and proved as Theorems 11 and 13 in my (1988). A probabilistic counterpart of the present claim, or rather a considerable generalization thereof, is proved as Theorem 5 in my (1980).
[44] Which is the basic theme of Putnam's recent work; cf., e.g., the introduction and ch. 4 of Putnam (1983).

[45] Proof: Let the i-proposition A be a direct cause of B in α^* relative to β^*. If $D = {}^{\alpha^*}(I_{<j}\text{-}\{i\})$, this says that $\beta^*(B \mid A \cap D) > \beta^*(B \mid \bar{A} \cap D)$. Let $E_1 = A \cap D$, $E_2 = \bar{A} \cap D$, $E_3 = A \cap \bar{D}$, $E_4 = \bar{A} \cap \bar{D}$, and $E = \bigcup\{E_r \mid \beta^*(B \mid E_r) \geqslant \beta^*(B \mid E_1)\}$. The law of disjunctive conditions (after Definition 3) immediately implies that $\beta^*(B \mid E) > \beta^*(B \mid \bar{E})$, i.e. that E is a (unconditional) reason for B relative to β^*. The same reasoning applies if B has a direct effect in α^* instead of a direct cause.

[46] This is so because, as the proof in the previous note shows, there is an unconditional reason for B, whenever there is a conditional reason for B.

[47] On this score, then, the sentential and Sneed's and Stegmüller's structuralist view of theories seem equally insufficient. This insufficiency is also felt, for instance, by Kitcher (1981), when he associates explanatory stores of argument patterns with scientific theories. Cf. also Mühlhölzer (1989), ch. 6.

[48] Of course, the demon has other epistemological defects as well. For instance, it may be one of the two gods of Lewis (1979), pp. 502f., unable to localize itself. But this is obviously another kind of defect.

[49] The proofs given in the notes 43 and 45 also apply to these claims.

[50] This means that Ω_i has only two elements. This premise is technically required and I am not sure about the best way to get rid of it.

[51] Proof: For each $\omega \in {}^{s}C_{A,B}$ we have $\kappa(\bar{B} \mid A \cap {}^{\omega}(I_{<j}\text{-}\{i\}) > 0$ and $\kappa(\bar{B} \mid \bar{A} \cap {}^{\omega}(I_{<j}\text{-}\{j\}) = 0$. Trivially, each $I_{<j}\text{-}\{i\}$-measurable $D \subseteq {}^{s}C_{A,B}$ is equal to $\bigcup\{{}^{\omega}(I_{<j}\text{-}\{i\} \mid \omega \in D\}$. Therefore, the law of disjunctive conditions (after Definition 3) implies the assertion that for each such D $\kappa(\bar{B} \mid A \cap D) > 0$ and $\kappa(\bar{B} \mid \bar{A} \cap D) = 0$.

[52] At this point it is particularly clear that my analysis of explanation is closely related to Hempel's requirement of maximal specificity (cf. Hempel (1965), pp. 397ff.) and to Skyrms' notion of resiliency (cf. Skyrms (1980) parts IA and IID).

[53] The failure of the analogous probabilistic assertion is related to Simpson's paradox. Cf. also my (1990a), pp. 128.

[54] Proof: Let ${}^{s}C_{A,B}$ be abbreviated by C. It was just shown that A is a sufficient reason for B conditional on C, i.e. that (a) $\kappa(\bar{B} \mid A \cap C) > 0$ and (b) $\kappa(\bar{B} \mid \bar{A} \cap C) = 0$. Since A causally explains B as necessary, A, B, and C are believed; this implies $\kappa(A) = 0$ and $\kappa(\bar{C}) > 0$; hence $\kappa(A \cap \bar{C}) > 0$ and (c) $\kappa(\bar{C} \mid A) = \kappa(A \cap \bar{C}) - \kappa(A) > 0$. And the additional premise says that (d) $\kappa(C \mid \bar{A}) = 0$. Now, (a) implies $\kappa(\bar{B} \cap C \mid A) > 0$, and (c) implies $\kappa(\bar{B} \cap \bar{C} \mid A) > 0$; therefore $\kappa(\bar{B} \mid A) > 0$. Moreover, (b) and (d) imply $\kappa(\bar{B} \cap C \mid \bar{A}) = 0$ and thus $\kappa(\bar{B} \mid \bar{A}) = 0$.

[55] Proof: Let ${}^{n}C_{A,B}$ be abbreviated by C. It was just shown that A is a necessary reason for B conditional on C, i.e. that (a) $\kappa(B \mid A \cap C) = 0$ and $\kappa(B \mid \bar{A} \cap C) > 0$, hence (b) $\kappa(B \cap C \mid \bar{A}) > 0$. Since A causally explains B as possible, A, B, and C are believed; thus $\kappa(A \cap C) = 0$ which implies (c) $\kappa(C \mid A) = 0$. And the additional premise says that (d) $\kappa(B \cap \bar{C} \mid \bar{A}) > 0$. Now, (a) and (c) imply $\kappa(B \cap C \mid A) = 0$ and thus $\kappa(B \mid A) = 0$. Moreover, (b) and (d) imply $\kappa(B \mid \bar{A}) > 0$.

[56] Hempel (1965), pp. 368. This is the part the thesis of the structural identity of explanation and prediction which Hempel (1965), pp. 364ff., endorses.

[57] I have in mind Michael Scriven's examples of the jealous murderer and the collapsing bridge discussed in Hempel (1965), pp. 370ff.

[58] For proof note that (a) $\kappa(\bar{B} \mid A) = \min[\kappa(\bar{B} \mid A \cap C) + \kappa(C \mid A), \kappa(\bar{B} \mid A \cap \bar{C}) + \kappa(\bar{C} \mid A)]$. We have assumed (b) $\kappa(\bar{B} \mid A) > 0$ and (c) $\kappa(\bar{B} \mid A \cap C) = 0$. All three

immediately imply the first claim $\kappa(C|A) > 0$. According to the law of negation (after Definition 1), the latter entails $\kappa(\overline{C}|A) = 0$; this and (a) in turn entail $\kappa(\overline{B}|A \cap \overline{C}) > 0$; and this and (c) say that C is a necessary reason for $\overline{B}$ given A.

[59] I refer to the observation in Savage (1954), sect. 7.3, that the expected utility of free information is always non-negative, and to the strong generalizations offered by Skyrms (1990), ch. 4. A different generalization to free memory may be found in Spohn (1978), sect. 4.4.

REFERENCES

Beauchamp, Tom, Rosenberg, Alexander: 1981, *Hume and the Problem of Causation*, Oxford University Press, Oxford.

Bunzl, Martin: 1979, 'Causal Overdetermination', *Journal of Philosophy* **76**, 134—150.

Cartwright, Nancy: 1979, 'Causal Laws and Effective Strategies', *Noûs* **13**, 419—437.

Cohen, Jonathan L.: 1977, *The Probable and the Provable*, Clarendon Press, Oxford.

Ellis, Brian: 1979, *Rational Belief Systems*, Blackwell, Oxford.

Field, Hartry: 1972, 'Tarski's Theory of Truth', *Journal of Philosophy* **69**, 347—375.

van Fraassen, Bas C. 1980, *The Scientific Image*, Oxford University Press, Oxford.

Friedman, Michael: 1974, 'Explanation and Scientific Understanding', *Journal of Philosophy* **71**, 5—19, German translation in Schurz (1988), pp. 171—191.

Gärdenfors, Peter: 1988, *Knowledge in Flux*, MIT Press, Cambridge, Mass.

Hempel, Carl Gustav: 1965, *Aspects of Scientific Explanation and Other Essays in the Philosophy of Science*, Free Press, New York.

Hume, David: 1739, *A Treatise Concerning Human Nature*. Page numbers refer to the 2nd edition by P. H. Nidditch of the edition of L. A. Selby-Bigge, Oxford University Press 1978, Oxford.

Hume, David: 1777, *An Enquiry Concerning Human Understanding*. Page numbers refer to the 3rd edition by P. H. Nidditch of the edition of L. A. Selby-Bigge of David Hume's Enquiries, Oxford University Press 1975, Oxford.

Hunter, Daniel: 1991, 'Maximum Entropy Updating and Conditionalization', this volume, pp. 45—57.

Kitcher, Philip: 1981, 'Explanatory Unification', *Philosophy of Science* **48**, 507—531, German translation in Schurz (1988), pp. 193—229.

Lambert, Karel: 1988, 'Prolegomenon zu einer Theorie des wissenschaftlichen Verstehens', in Schurz (1988), pp. 299—319.

Lambert, Karel: 1991, 'On Whether an Answer to a Why-Question is an Explanation if and only if it Yields Scientific Understanding', in: G. G. Brittan jr. (ed.), *Causality, Method, and Modality*, Kluwer, Dordrecht, pp. 125—142.

Lewis, David: 1973, 'Causation', *Journal of Philosophy* **70**, 556—567; also in Lewis (1986), pp. 159—172.

Lewis, David: 1979, 'Attitudes de dicto and de se', *Philosophical Review* **88**, 513—543.

Lewis, David: 1980, 'A Subjectivist's Guide to Objective Chance', in R. C. Jeffrey (ed.), *Studies in Inductive Logic and Probability*, vol. II, University of California Press, Los Angeles, pp. 263—293; also in Lewis (1986), pp. 83—113.

Lewis, David: 1986, *Philosophical Papers*, Vol. II, Oxford University Press, Oxford.
Mackie, John L.: 1974, *The Cement of Universe*, Oxford University Press, Oxford.
Mühlhölzer, Felix: 1989, *Objektivität und Erkenntnisfortschritt. Eine Antwort auf Thomas S. Kuhn*, Habilitationsschrift, Munich.
Putnam, Hilary: 1983, *Realism and Reason, Philosophical Papers*, Vol. 3, Cambridge University Press, Cambridge.
Rescher, Nicholas: 1973, *The Coherence Theory of Truth*, Oxford University Press, Oxford.
Rescher, Nicholas: 1976, *Plausible Reasoning*, Van Gorcum, Assen.
Rescher, Nicholas: 1985, 'Truth as Ideal Coherence', *Review of Metaphysics* **38**, 795—806.
Salmon, Wesley C.: 1970, 'Statistical Explanation', in: R. G. Colodny (ed.), *Nature and Function of Scientific Theories*, University of Pittsburgh Press, Pittsburgh, pp. 173—231.
Salmon, Wesley C.: 1978, 'Why Ask, 'Why?'? An Inquiry Concerning Scientific Explanation', *Proceedings and Addresses of the American Philosophical Association* **51**, 683—705.
Salmon, Wesley C.: 1984, *Scientific Explanation and the Causal Structure of the World*, Princeton University Press, Princeton.
Savage, Leonard J.: 1954, *The Foundations of Statistics*, Wiley, New York.
Schurz, Gerhard (ed.): 1988, *Erklären und Verstehen in der Wissenschaft*, Oldenburg, Munich.
Shackle, G. L. S.: 1961, *Decision, Order, And Time in Human Affairs*, Cambridge University Press, Cambridge.
Shenoy, Prakash: 1991, 'On Spohn's Rule for Revision of Beliefs', *International Journal of Approximate Reasoning* **5**, 149—181.
Skyrms, Brian: 1980, *Causal Necessity*, Yale University Press, New Haven.
Skyrms, Brian: 1990, *The Dynamics of Rational Deliberation*, Harvard University Press, Cambridge, Mass.
Spohn, Wolfgang: 1978, *Grundlagen der Entscheidungstheorie*, Scriptor, Kronberg/Ts.
Spohn, Wolfgang: 1980, 'Stochastic Independence, Causal Independence, and Shieldability', *Journal of Philosophical Logic* **9**, 73—99.
Spohn, Wolfgang: 1983, 'Deterministic and Probabilistic Reasons and Causes', in C. G. Hempel, H. Putnam, W. K. Essler (eds.), *Methodology, Epistemology, and Philosophy of Science. Essays in Honour of Wolfgang Stegmüller on the Occasion of his 60th Birthday*, Reidel, Dordrecht, pp. 371—396.
Spohn, Wolfgang: 1988, 'Ordinal Conditional Functions. A Dynamic Theory of Epistemic States', in: W. L. Harper, B. Skyrms (eds.), *Causation in Decision, Belief Change, and Statistics*, Vol. II, Kluwer, Dordrecht, pp. 105—134.
Spohn, Wolfgang: 1990a, 'Direct and Indirect Causes', *Topoi* **9**, 125—145.
Spohn, Wolfgang: 1990b, 'A General Non-Probabilistic Theory of Inductive Reasoning', in: R. D. Shachter, T. S. Levitt, L. N. Kanal, J. F. Lemmer (eds.), *Uncertainty in Artificial Intelligence* **4**, Elsevier, Amsterdam, pp. 149—158.
Spohn, Wolfgang: 1992, 'Causal Laws are Objectifications of Inductive Schemes', in: J. Dubucs (ed.), *Theory of Probability*, Kluwer, Dordrecht, to appear.
Teller, Paul: 1976, 'Conditionalization, Observation, and Change of Preference', in: W.

L. Harper, C. A. Hooker (eds.), *Foundations of Probability Theory, Statistical Inference, and Statistical Theories of Science*, Vol. I, Reidel, Dordrecht, pp. 205–259.

Tooley, Michael: 1987, *Causation*, Oxford University Press, Oxford.

JULES VUILLEMIN

THE SYSTEMS OF PLATO AND ARISTOTLE COMPARED AS TO THEIR CONTRIBUTIONS TO PHYSICS

The science of kinematics did not arise before Christian Huyghens was able to analyse circular motion, though this analysis was implicit in Descartes' reflections and, from the time of Eudoxus and especially Ptolemy on, ingenious techniques concerning the measurement of superimposed or deformed circular motions had been developped in celestial kinematics. As to dynamics, if the law of falling bodies is due to Galileo and the principle of inertia is conceived in its generality by Descartes, the principles of motion find their first systematic expression in Newton's *Principia*. Newton's principles and their development during the 19th century gave to the relation between kinematics and dynamics a form which has been questioned by quantum mechanics. I aim to seek in the philosophical analysis of the concept of motion by Plato and Aristotle some explanations of the difficulties discovered in the relations between kinematics and dynamics and more generally in the history of physics.

I. A COMPARISON BETWEEN PLATO'S AND ARISTOTLE'S CONCEPTS OF MOTION

It is difficult to definitely assess Plato's proper concept of motion because of the impossibility of making a clear distinction between the statements advocated by the protagonists of the Dialogues and Plato's own theory. Moreover the genuine doctrine is often revealed and concealed in the guise of mythical expression. Nevertheless the following comparison may safely be drawn between the two greatest philosophers of Antiquity:

	PLATO	ARISTOTLE
I	1. The first origin of any motion or change is selfmotion, called also soul or life.	1. There is no self motion. Every motion is *ab alio*.

Wolfgang Spohn et al. (eds.), Existence and Explanation, 197–206.

	2. Being without external cause, selfmotion is undestructible and perpetual (immortality of the soul).	2. If there exists a perpetual motion, it is caused and its perpetuity is due to the sempiternality of the action of its cause.
	3. Selfmotion precludes in itself any distinction between mover and moved. or action and passion.	3. Every motion results from the affection of a moved by a mover.
	4. Selfmotion is a purely spiritual motion without relation to space or to any motion which proceeds in space and is a possible object of sensation. Time is an image of eternity.	4. Every motion is sensible. Its necessary condition is a trajectory in space during a given time. Time measures motion.
	5. Selfmotion is pure actuality.	5. Every motion is the actuality of a potency quâ potency.
	6. Kinematics being the science that describes bodily and material motion in space, spiritual dynamics is not in the least degree dependent on kinematics.	6. Kinematics is the necessary condition of dynamics; all dynamics is subject to the laws of kinematics.
II	1. There is no adequate sensible representation of selfmotion such as the mind's. It is the apprehension of a form or idea.	1. Every motion expresses a potency, unintelligible without a privation somehow felt by matter aspiring to its form.
	2. Its least imperfect image would be axial rotation ("an in one place move in the same and like manner, in and about the same, and in relation to the same, and according to one proportion and order") (*Laws*, X 897a–b).	2. The most perfect motion is produced by a substance whose matter reaches its least degree and is reduced to place (sempiternal sensible substances).
	3. Selfmotion immediately produces communicated motion.	3. Prime motion is pure and sempiternal locomotion.
	4. We do not know the nature of mind's efficiency and we may frame various hypotheses.	4. Sempiternal locomotion is nothing else than the restlessness in beings striving to imitate the immobility of the prime mover.

5. By producing communicated motions spiritual selfmotion informs the primeval chaos into a sensible world that imitates the ideas as far as this is possible. The primeval chaos is not a prime matter, but the immediate, instable and fuzzy whole of the imitations of the ideas by the receptacle, *i.e.* the completely unordered appearances into which the indefinite extension or spatiality makes sensible their first imitations. Physical motions are thus ordered into a hierarchy of perfections which reflects the hierarchy of the perfections of the knowing souls and the hierarchy of the regularities of natural laws.

5. Local matter accounts for a motion without contraries insofar as pure elsewhere excludes any local contrariety (axial rotation) Full matter is relevant to motion from a contrary to a contrary and therefore to a motion limited in time. Motions can be ordered according to a hierarchy of perfections depending on the relative success of form in informing matter.

6. Mechanisms are secondary physical motions induced by the relations that first communicated motions have with otherwise organized bodies. All motions resort to the calculations of more or less perfect souls. Their plurality insofar as it introduces a last element of disorder and irregularity shows that the transformations of appearances into phenomena leaves an irrational residue of approximation, characteristic of the sensible world.

6. Two concurring motions produce a resulting mechanism. The generical and numerical plurality of enmattered substances explains contingency in the world, which is there in the last resort because prime matter cannot be completely actualized.

7. Physics as the theory of motion describes how final causes, produced by souls copying ideas, are superimposed on errant or mechanical causes and approximately govern them. This organization of the sensible world would not be accountable without the activity of the best souls of the gods.

7. There is only one immobile Prime mover who, without any activity of Himself (except His own reflection) and only because of His final perfection, moves the Heavens and the Earth and makes matter strive after its form.

	8. Insofar as the sensible world only approximately imitates the ideas, physics is intended to 'save the phenomena'.	8. There is no world aside from sensible world. Physics describes reality.
III	1. The first of all motions is self-motion which defines soul and is the contemplation of the ideas.	1. The *noèsis noèseôs*, selfcontemplation of the Prime mover, is no motion.
	2. All communicated motions depend on selfmotion and order with more or less perfection the participation of the sensible world in the ideas.	2. All motion is sensible and, expressing the actualization of form in matter, depends on the Prime mover as final cause.
	3. The perfection of a sensible motion depends in its being more or less akin to spiritual motion.	3. The perfection of a motion depends on the kind of matter involved in it, which is determined by the category to which it is assigned. Motion in general therefore is not a category but a transcendantal.
	4. The most perfect sensible motions are those by which the receptacle is ordered into space (the production of the three dimensions).	4. The most perfect motion is the motion involving merely the topic hylè. Being only local and foreign to any contrariety, this motion must be continuous, uniform (with no change in quantity and quality) and never at rest (with no change in the substantial mover). Here arises the *kinematical* principle of the primacy, simplicity and perfection of circular uniform motion.
	5. Among translations (φορά), the motion which is the most akin to spiritual motion is axial motion. Resting at the centre the motion takes place in 'one place' and its swiftness is proportionally distributed to greater and smaller circles ($v = R\omega$, ω being the angular velocity) (*Laws*, 893).	5. Though apparently rectilinear motions only involve local matter in their definition, the theory of natural places prevents them from being sempiternal and continuous as well as from being always natural (they must come to a rest when the mobile has reached its natural place, and they must be violent when it leaves it). Therefore they cannot be always uniform. Therefore they are involved in a change of quality (alloiôsis). Therefore straight inertial motions

	are impossible. This causes a complete separation of the sensible world into the Heavens and the sublunar, properly material world.
6. After axial motion comes rolling. The effect of rolling may be illustrated by Eudoxus' hyppopedes in astronomy. There does not seem to be a fundamental distinction between this celestial type of change of quality and the terrestrial ones (aging or whitening, *Theaitetus*, 180c—181b).	6. Augmentation and diminution involve quantity, while alteration involves quality.
7. The last communicated motions are agregations and disgregations.	7. The motions the most involved with matter are generation and corruption, which produce the birth and death of substances.

II. CONSEQUENCES OF THESE PRINCIPLES FOR THE HISTORY OF PHYSICS

Thus there are important differences between Plato's and Aristotle's theories of motion. Spiritualism dissociates dynamics from kinematics, while hylemorphism results in subordinating all dynamical manifestations to kinematics. The theology of the immovable Prime Mover does not produce the same consequences as the theology of selfmoving gods. Aristotle cuts the unity of the sensible world into two separate realms: the lawlike celestial world where everything happens with necessity and the sublunar world where motion is submitted to contrariety and where all events are tinged with contingency. Plato's opposition is between ideas and their sensible imitations and it allows him to save the unity of the sensible world which, even in celestial motion, is characterized by its being an approximation and always falling short of its pattern. Against Aristotle's dualist realism in physics, the Platonists will maintain the unity of physics but will also limit its ambition to saving the phenomena. Finally, Aristotle binds motion to potentiality. Though a mathematical analysis of motion in terms of instantaneous velocity and acceleration is not *ipso facto* prohibited on this conception, it is at least hindered since such an analysis runs against the ontology of temporal development. Plato, for whom motion is pure actuality, is not hampered by this difficulty.

Despite all these differences, both systems rely on the strictest dependence of physics on theology. This dependence will have far reaching consequences for the development of physics. Four of them have been of the utmost importance.

1. Both systems maintain the contrariety between motion and rest. No principle of inertia is therefore thinkable (Plato's selfmotion is only spiritual). Descartes will be the first to clearly state this principle in terms of rectilinear motion, a principle fully expressed by Newton's first law of motion.

2. Even Plato grants uniform circular motion a perfection transformed by Aristotle into a kinematical principle, according to which uniform circular motion is kinematically simple. It is remarkable that Galileo himself will maintain this erroneous conception.[1] Only Huyghens will know how to analyse circular motion in mathematical terms, an analysis which was in principle not beyond Descartes' rèach, but was prevented by his representation of the vacuum.

3. Eudoxus' system of homocentric astronomy may be considered as the main scientific consequence of Plato's or Aristotle's principles. Since Eudoxus himself adopted a third philosophical view (the ideas are in the sensible), such a notion of 'scientific consequence' deserves analysis. We know that it is impossible to account for the motions of the bodies belonging to our solar system either by composing axial rotations according to Eudoxus' method or even by renouncing homocentricity and maintaining a composition of excentred uniform circular motions as Ptolemy, Copernicus and Tycho-Brahé did. Nevertheless, although this last general hypothesis must be recognized as false, it has permitted a good approximation of many celestial motions. Moreover it is highly probable that, had the Greek geometers not tempted their kinematical constructions from this simple and false, but precise and fortunately approximate hypothesis, they should never have founded celestial kinematics.

A false but productive hypothesis in science is better then no hypothesis at all. However the methodology of science cannot be content with this assertion, though many astronomers have interpreted Plato's program as meaning that the truth or the falsity of a hypothesis does not matter as long as it saves the phenomena. Kepler,[2] reflecting on the so called equivalence of the astronomical hypotheses (Ptolemy,

Tycho, Copernicus) adopts a better principle. When observed facts are consequences of a false hypothesis, he says, the relation of consequence is only an accidental one. The implication becomes necessary (according to the principle of the syllogism of the cause) only when the hypothesis is true. In other terms, the approximately true description of the motions of the bodies of our solar system has no connection with the principle according to which uniform circular motions are simple nor with the metaphysical principles of Plato's or of Aristotle's physics. This success depends on two different facts: a) a fact relevant to kinematics, namely the slightness of the eccentricities of the elliptical motions within the solar system, b) a fact relevant to dynamics, namely the overwhelming superiority of the mass of the sun.

4. The theological cause of motion is in Plato a formal one, in Aristotle a final one. Neither of them is in itself material or efficient. This implies that mobiles, insofar as they move for realizing the order of the ideas or for actualizing their own potnetiality, do not exert any counteraction upon their immaterial cause. This cause remains beyond passion and action. Or if we insist on calling action either formal or final causality, this action has no counterpart of passion. Therefore even when the principle of dynamics depends on local kinematics, as is the case with Aristotle, it does not fall under the categories of action and passion. It will not be until Newton introduces with his second principle the concept of mass into the concept of force and formulates with his third principle the equality of action and reaction, that the remmants of Greek immaterial dynamics will have disappeared.

III. HOW TO CONCEIVE OF SENSIBLE KNOWLEDGE?

A further principle, implicitly implied by the definition of measurement, is also involved in the discussion. It is connected with logic as the theory of relations, with epistemology and psychology, with physics as the theory of motion and with metaphysics and theology as the theory of causality, but it is rarely analyzed as regards these four connections. The principle is required if the statements of physics are to be recognized some objectivity, *i.e.* if Protagoras' relativism is not to be accepted as the last word of human wisdom. According to Protagoras, man is the maesure of all things. As understood by Plato in the *Theaitetus*, this doctrine amounts to two theses: 1) all knowledge is sensation, 2) sensation results from an instantaneous meeting of the motion of the sentient with the motion of the sensed. There is strictly speaking no being, but only happenings. Protagoras does not question realism: he does question objectivity, since these happenings are relative to the individuals which affect them and are affected by them; therefore they belong to strictly 'private worlds': "my perception is inerrant, for it is relative to my world (*ἐμὴ οὐσιὰ*) at that moment" (160c).

Although Plato criticizes and dismisses Protagoras' theory, he does not state positively how we have to conceive of sensible truth. We cannot content ourselves with identifying sensation with knowledge, nor being with universal motion, since nothing definite could then be asserted and Protagoras' thesis is selfdefeating. Sensation needs knowledge just as motion needs rest and ideas. We learn in the *Timaeus* (45c) that vision is a rectilinear motion, whereas thought is akin to the divine axial motion. But except for saying that sensation occasions thought and reflection, Plato leaves open the status of sensible measurement.

Its determination is one of the merits of Aristotle who aims at justifying realism. If we simplify the clouded and often perplexing doctrine, it seems that, among all the relations, Aristotle isolates and puts into a same group a) epistemological relations between what measures, knows and thinks and what is actually measured, known and thought b) relations of action and passion insofar as they actually obtain. He opposes this group to ordinary relations such as 'numerical' relations (*e.g.* double/half). The relations which belong to this second group are such that both relata are what they are in virtue of their relations obtaining. For example, if a is a magnitude, there exists another magnitude b such that a is greater than b, and reciprocally, if b is a magnitude, there exists another magnitude a such that b has to a the converse relation of being lesser than it. It is that reciprocation which does not always hold for the relations of the first group. If a is a mathematician, there exists a proposition p such that p is a mathematical truth such that a knows p, but it is not always the case that, if p is a mathematical truth, there exists a mathematician a such that p is known by a. In the same way, according to Aristotle, and however unintuitive this statement may be for us, if a is moved, there exists a mover b such that a is moved by b, but it is not always the case that if b is a mover, there exists a moved a such that b moves a.[3] It is remarkable that Aristotle extends without discussion to sensible knowledge the realism he accepts for intellectual knowledge.

Relations of the first group, *i.e. relationes modo intelligentiae* and relations of passion/action, are therefore characterized by a kind of asymmetry of what from the status of one relation can be inferred for the other. Knowledge, measurement, passion have an epistemological priority and allow an ontological inference, while the known, the measured, the action have an ontological priority which, except if particular circumstances prevail, blocks any epistemological inference.

From a psychological point of view, Protagoras' relativism results from a confusion between epistemological and ontological priority. It is refuted as soon as one agrees that what is prior for knowledge is posterior for ontology and *vice versa*. Let us observe that this ontological priority of being over being known would be destroyed if measurement would by itself operate on the state of the being (a state which according to the situation we conceive of as rest or motion) and disturb it. To prevent this possibility, sensible measurement, knowledge and thought are defined as passions, but these passions in their potentiality are receptions of forms without matter and in their actuality receptions of ratios between sensible magnitudes rather than receptions of these magnitudes themselves (which possibly would destroy the organs (*De anima*, II 12)). Aristotle hence accepts the following principle: if, if a is the result of an observation, there exists an observable state of affairs b such that a measures b, it is not for that, that if b is an observable state of affairs, there exists a result of observation a such that b is measured by a.

The same remarks apply to physics, because of the logical analogy between the relation of measurement and the relation of motion. We take notice of the existence of motions (and among them of sempiternal celestial motions). Therefore we must posit a Prime Mover. On the one hand, the last dynamical cause and magnitude cannot be grasped by our sense. On the other hand, any material change necessarly expresses itself through kinematical magnitudes which impinge upon our senses through their form only, that is, without disturbing the perceived motion.

Finally the theology which is needed by physics is, according to the different priorities of epistemology and ontology, an *a posteriori* theology which admits only of proofs 'by the effects'. Effects are celestial motions or, indirectly, motions by themselves. The principle of inertia at once eliminates these proofs. When Newton calls on the Lord, it will be for a completely other reason, namely the improbable distribution of heavenly motions, though his concepts of absolute time, space and motion still imply a reference to God's 'organs' of knowledge.

As to the conception of sensible knowledge as identified with the reception of a form, it was already denied in Antiquity by all materialist systems. It is remarkable however that this denial remained without empirical consequence in physics until the advent of quantum mechanics.

Les Fourgs, France

NOTES

[1] M. Clavelin, *La philosophie naturelle de Galilée*, A. Colin, Paris, 1968, pp. 252—253, p. 373.

[2] J. Kepler, *Mysterium Cosmographicum*, 2° ed. (1621), M. Caspar (ed.) (*Gesammelte Werke*, München, VIII 31 C^1—33 C^2).

[3] On the question and on the real or alleged logical difficulties of the Aristotelian expression of realism, see J. Vuillemin, *De la logique à la théologie, cinq études sur Aristote*, Flammarion, Paris, 1967, quatrième étude, pp. 147—163, and the criticism of J. Brunschvicg, 'Le Dieu d'Aristote au tribunal de la logique', *L'âge de la science*, Dunod, Paris, III n° 4 (oct.—déc. 1970). pp. 323—343. The present summary or rather simplification is disentangled from these difficulties.

PAUL WEINGARTNER*

A NOTE ON ARISTOTLE'S THEORY OF DEFINITION AND SCIENTIFIC EXPLANATION

The purpose of the paper is to show (1) that Aristotle had a kind of criterion of non-creativity in his theory of definition and (2) that he had a requirement for scientific explanations which amounts to finding an interpolation-sentence. The relevant passages can be found in the *Posterior Analytics*, second book, especially chapters 2, 7, 8, 13.

Moreover it will soon be clear from the article that Aristotle's theory of definition includes an important principle which was — 2200 years later — stressed by Meinong and which Lambert called the Principle of Independence.[1] A further purpose of the article is therefore to show to my friend Joe (Lambert) that already Aristotle used this principle as a very important part of his theory of definition.

1. ARISTOTLE'S CRITERION OF NON-CREATIVITY

1.1. *General Characteristics of Definitions According to Aristotle*

1.11 Every scientific problem can be formulated in one of the following four forms: Is it the case that *A* is *B*? Is *A* the case? (Or: Does *A* exist?) Why is *A B*? What is *A*? According to Aristotle we are searching for an answer by means of a proof, i. e. a syllogism. Keeping this in mind the first two questions can be interpreted as being of the form: Is there a middle term? That is to ask whether *A* is *B* or whether *A* exists is to ask whether there is a middle term to account for *A*'s being *B*, or for *A*'s existing. The last two questions on the other hand can be interpreted as being of the form: What is this middle term? That is to ask why *A* is *B* or what *A* is, is to ask what the middle term is.[2]

1.12 A definition is not a demonstration (proof) or a scientific explanation. The reasons he gives are the following: (1) There are demonstrations which are not definitions. This is proved by the fact that every definition is universal and affirmative but there are demonstrations with the help of syllogisms which include negative or particular premisses. But it holds also that (2) there are definitions which cannot

Wolfgang Spohn et al. (eds.), Existence and Explanation, 207—217.

be demonstrations. Since there are some definitions which are first principles; but first principles cannot be demonstrated. (1) and (2) say that not every demonstration is a definition and not every definition is a demonstration.

According to Aristotle even the stronger statement holds that nothing is both a definition and a demonstration; i.e. 'definition' and 'demonstration' are incompatible concepts. The reason he gives for that is that a definition makes clear *what* something is whereas a proof or demonstration says *that* something is. A further reason for keeping separate essence and existence is according to Aristotle — that there is nothing in this world such that its existence would belong to its essence, since all things (except God) exist contingently.[3]

In respect to what has been said in chapter 1.11 the first two questions or its Aristotelian interpretation 'Is there a middle term?' are answered by a demonstration. Whereas the last two questions or its Aristotelian interpretation 'What is the middle term?' are answered by a definition.

1.13 A definition (answering a 'what is?'-question) cannot be the conclusion of a demonstration (of a syllogism). The reasons he gives are the following: Every definition is a universal convertible statement (i.e. an equivalence or an identity) answering a what-question. If the conclusion of a syllogism is convertible and universal then so must be both premisses. But in this case the middle term is already introduced by a definition in the premisses such that the whole demonstration is a petitio principii. On the other hand if the premisses are not convertible (and universal) then the conclusion cannot be a definition.[4]

1.4 In chapter 5 Aristotle says that a definition cannot be obtained by making a division. Since there is no necessity for such or such a division whereas a definition which answers a 'what-is-question' requires necessity.[5]

1.2. *The Criterion of Existential Non-creativity*

1.21 The most explicit passages for Aristotle's criterion of existential non-creativity are to be found in chapter 7 of the second book of the *Posterior Analytics*. In the first part of chapter 7[6] Aristotle gives a summary of the forgoing chapters saying that there is no way that definitions which answer 'what is-questions' could be demonstrated as conclusions of syllogisms. In the second part[7] he tries to answer the question whether the essence (the 'what is') can be achieved by demonstration or by definition. First he states that a demonstration necessarily states *that something is*; it is the business of demonstration to prove existence. But from such a proof of existence the essence of something could only follow if for at least one thing its being or existence would be the same as its essence. But for no thing, Aristotle says, its existence is its essence.[8] Thus since the business of demonstration is to prove the existence it cannot be the business of definition to do so.[9] The important passage seems to be the following one:

> It is clear, if we consider the methods of definition now in use, that those who define do not prove existence. Even if there is a line equidistant from the centre, *why* does that which has been thus defined exist? and *why* is this the circle? One might just as well call it the definition of mountain-copper. For definitions do not show either that the thing mentioned in the definitory formula can exist, or that it is that of which they claim to be definitions; it is always possible to ask why.[10]

1.22 From the passage cited in 1.21 it seems clear that for Aristotle a definition is not an existence assumption. It states *what* something is but not *that* it is or *exists*. And as he says from a definition of a circle it does not follow that circles exist. Thus no existential statements or existence assumptions about the thing defined or about some part of the definiens can be derived from a definition. If this is correct one may ask whether Aristotle requires definitions to be non-creative in the sense of modern logic: i.e. in the sense that no new theorem (not derivable without the definition) in which the defined term does not occur can be derived from the definition. But this cannot be the case. The reason is that Aristotle understands definitions which state the essence of things or which answer 'what-questions' at least sometimes as principles from which other theorems can be derived.[11] Therefore his criterion of non-creativity must be weaker than the criterion of non-creativity used in modern logic. It allows to derive new theorems (previously unprovable) from the definitions provided they are not existence assumptions of the definiendum or of a part of the definiens.

1.23 Aristotle's criterion can be formulated in a more exact way by 1.231 or 1.232:

1.231 A definition satisfies the Aristotelean criterion of (existential) non-creativity iff it does not permit the proof of existence theorems about the definiens or the definiendum (or parts of it) which were previously unprovable.

1.232 A definition D introducing a new symbol of a theory satisfies the Aristotelean criterion of (existential) non-creativity iff there is no formula T which affirms the existence of the definiendum or of the definiens or of any part of it such that $D \rightarrow T$ is derivable from the axioms and preceding definitions of the theory but T is not so derivable.

1.24 *Additional Remarks and Examples*

(1) Aristotle when speaking of definitions has in mind only explicit definitions.

(2) According to Aristotle individuals in the proper sense, i.e. substances like Socrates, Plato, Gorbatschov or the horse Nelli cannot be defined.[12]

(3) Definitions in the proper sense, i.e. such which state the essence or which answer 'what-questions' are concerned with properties (resp. classes) of first order (species).

(4) Definitions are true equivalences or identities.[13]

From (1) to (4) it is clear that a definition like

> x is an animal $\leftrightarrow$ x is a living being $\wedge$ x has senses or
> x is a living being $\leftrightarrow$ x has growth $\wedge$ x has nutrition $\wedge$ x has propagation

is less close to the Aristotelian view than its explicite versions:

> animal = living being $\cap$ being with senses
> living being = being with growth $\cap$ being with nutrition $\cap$ being with propagation
> men = rational being $\cap$ animal.

Nevertheless if a definition occurs in a demonstration (cf. 2.1 and 2.2) both definiendum and definiens can be stated in sentences which are connected by equivalence (instead of being stated as classes connected by identity).

In respect to these explicit examples it is easy to say what Aristotle's criterion of non-creativity forbids: It forbids the derivation of statements like: there are living beings, there are beings which have senses, there are rational beings, there are beings which have propagation . . . etc.

1.25 Application of Aristotle's criterion of non-creativity to modern logic.

Since Aristotle has in mind only explicit definitions the versions of his criterion given in 1.23 need not to contain a condition which says that the existence theorem about the definiens has to be stated in primitive terms or previously defined terms; i. e. the newly introduced term cannot occur in the definiens if the definition is explicit.

But since this assumption is not always made in modern logic one has to revise the formulations given in 1.23:

1.251 A definition satisfies the Aristotelean criterion of (existential) non-creativity iff it does not permit the proof of existence theorems stated in primitive terms (or previously defined terms) about the definiens (or parts of it) which were previously unprovable.

1.252 A definition *D* introducing a new symbol of a theory satisfies the Aristotelean criterion of (existential) non-creativity iff there is no formula *T* which affirms the existence of something and in which the new (introduced) term does not occur, such that $D \rightarrow T$ is derivable from the axioms and preceding definitions of the theory but *T* is not so derivable.

A serious question here is what kind of existence-assumptions can make a definition existentially creative. Since individuals cannot be defined according to Aristotle it seems that he would forbid only such theorems which state the existence of properties or classes. But on the other hand he says that only demonstrations prove the existence of something (including individuals) whereas definitions do not prove the existence of anything but leave existence-questions open. If the latter is correct then both existence-assumptions concerning individuals and existence-assumptions concerning properties or classes must not be derivable from definitions.

1.26 Example

Examples of definitions which are creative (in the sense of modern logic) but do not make any existence-assumption occur in Lesniewski's system of Protothetic. Consider as an example the following axiom for the classical propositional calculus:

CCCNuvwCCCCCpqCNrNsrtCCtpCsp

If the rules of substitution and detachment are allowed only the rule of substitution can be applied since one has not got any antecedent for applying the rule of detachment. However, if the following definition of alteration is introduced detachment is applicable and one can derive at once Meredith's axiom for the full classical propositional calculus:[14]

CCNuvAuv CAuvCNuv	} Definition

If one substitutes the newly introduced expression 'Auv' for 'w' in the above axiom one can detach Meredith's axiom, where the first part of the above definition functions as the antecedent.

This shows that the above definition is creative in the modern sense of the word. But it is not existentially creative since it does not allow any existence-theorems to be derived. Thus such definitions are allowed by the Aristotelean criterion of (existential) non-creativity.

2. ARISTOTLE'S REQUIREMENT FOR AN INTERPOLATION SENTENCE

2.1 As already stated according to Aristotle a definition cannot occur as a conclusion of a demonstration (cf. 1). Nevertheless one way to find a definition which answers a 'what-question' is to find a middle term which can serve as the definiens such that the respective definition stating the essence is a premiss in the syllogism.[15]

Furthermore the answer to a 'what-question' is according to Aristotle in many cases also an answer to a 'why-question', i.e. is to find a cause which is the aim of a scientific explanation.[16] Therefore to give an answer to a scientific problem means to find a middle term which can serve as a definiens of a definition answering a 'what-question' and a 'why-question' in a syllogism.

2.2 To search for a middle term in a syllogism (demonstration) such that this middle term is the definiens of a definition stating the essence is nothing but to search for an interpolation sentence. The reasons for that are the following:

(1) Every scientific explanation has to be stated in the first figure;[17] i.e. in one of the following syllogisms:

AAA, EAE, AII, EIO.

(2) Knowledge of the essence and knowledge of the cause have to be positive and universal according to Aristotle. Moreover the cause must occur as the middle term in the syllogism.[18]

(3) From (1) and (2) and from the fact that definitions are always universal and affirmative it follows that the only syllogism suitable for a scientific explanation which states the essence and the cause is the modus AAA (BARBARA).

(4) The modus BARBARA reinterpreted in Predicate Logic of First order has the form:

$$(\forall x)(Mx \rightarrow Px), (\forall x)(Sx \rightarrow Mx) \vdash (\forall x)(Sx \rightarrow Px)$$

Without bound variables this is the form of of a hypothetical syllogism with converted premisses:

$$B \rightarrow C, A \rightarrow B \vdash A \rightarrow C.$$

(5) To search for the middle term which functions as the definiens of a definition means to search for the interpolation-sentence Mx (or: B) which connects Sx and Px and which contains the essential cause M. The major premiss in this case is a definition of the form $(\forall x)(Px \leftrightarrow Mx)$. The fact that definitions stating the essence and explanations stating the cause are always necessary truths (not just factual) according to Aristotle[19] need not concern us here.

(6) Whereas Craig has proved that, given $A \rightarrow C$, there is always a sentence B such that it holds that $A \rightarrow B$ and $B \rightarrow C$ (where B contains only predicates occuring in both A and C), Aristotle has required to find such a sentence B or better Mx which introduces a new predicate — the middle term — in order to give a sound scientific explanation, i.e. an explanation which states the essential cause.

(7) What has been said so far may be illustrated by the following example:

1. Eclipse is (defined as) the interposition of a heavenly body of the solar system between the body (which is in eclipse) and the sun. Or more accurately: x is in eclipse $=_{df}$ $(\exists y)$ y is a heavenly body of the solar system and y is between x and the sun. Symbolically: $(\forall x)[ECx \leftrightarrow (\exists y)(HBy \wedge Byxs)]$. Reduced to a syllogistic premiss: $(\forall x)(Px \leftrightarrow Mx)$ where 'M' is a complex predicate and 'P' stands for 'EC'.

2. If a heavenly body x (of the solar system, for example the moon) is in shadow then there is another heavenly body of the solar system y which is between x and the sun. Symbolically: $\forall x[(HBx \wedge SHx) \rightarrow$

$(\exists y)$ $(HBy \wedge Byxs)$]. Reduced to a syllogistic premiss: $(\forall x)$ $(Sx \rightarrow Mx)$ where 'S' is a complex predicate and 'M' is as above.

3. Therefore: If a heavenly body x (of the solar system, for example the moon) is in shadow then x is in eclipse. Symbolically: $(\forall x)$ $[(HBx \wedge SHx) \rightarrow ECx]$. Reduced to the syllogistic conclusion: (x) $(Sx \rightarrow Px)$.[20]

(8) From (1) to (7) it follows: Aristotle's requirement for one form of a genuine scientific explanation is the requirement to find a middle term M in a syllogism (more accurately in a modus BARBARA) which is at the same time a definiens of a (Aristotelian) definition answering a 'what-question' i.e. 'stating the essence of something' and answering a 'why-question'. Because of the facts stated in (1) to (7) and especially because only BARBARA is accepted as a demonstration of this kind the requirement amounts to finding an interpolation sencence Mx. This interpolation sentence satisfies Craig's interpolation theorem 'given $A \rightarrow C$ there is always a sentence B such that $A \rightarrow B$ and $B \rightarrow C$' except the condition 'where B contains only predicates occurring in both A and C'. Thus Aristotle's description of one form of a scientific explanation is this: A scientific explanation is a demonstration of the form modus BARBARA where the major premiss is a definition with the definiens as the middle term (or more accurately: where the definiens is the sentence containing the middle term, (cf. (7) 1.). In this case to ask why S (subject term) is P (predicate term) — or why $(\forall x)$ $(Sx \rightarrow Px)$, in the example: why any heavenly body of the solar system in shadow is in eclipse — is to ask whether there is an M such that $(\forall x)$ $(Mx \leftrightarrow Px)$ and (x) $(Sx \rightarrow Mx)$, i. e. whether there is an interpolation sentence Mx. But to ask whether there is an Mx is to ask if there is a reason (cause) for Px or is to ask why does Px occur; in the example: a reason (cause) for a heavenly body x to be in eclipse. For Mx (or M) is the genuine reason or essential cause for Px (for P) since it is at the same time the definiens for Px (for P) such that we really have both: an answer to a 'what-question' and an answer to a 'why-question'.

2.3. Historical remarks

Already in 1949 W. D. Ross realizes the essential structure of a scientific explanation according to Aristotle. He writes:

If, starting with a subject C and an attribute (or event) A, we can establish a connexion between C and A by a series of intermediate propositions such as 'that which has the attribute B_1 necessarily has the attribute A, that which has the attribute B_2 necessarily has the attribute B_1 ... that which has the attribute B_n necessarily has the attribute

B_{n-1}, C necessarily has the attribute B_n', then we know both that and why C has the attribute A.[21]

He could not connect this with Craig's Interpolation-Theorem since Craig's paper was published in 1957.[22]

Beth, however, refers to Aristotle when discussing Craig's interpolation-theorem and his own definability-theorem:

It seems only fair also to mention an older logician, namely Aristotle, because of his constant emphasis on the importance of the middle term in the syllogism. This conception has sometimes be criticized by representatives of mathematical logic. It is of course out of the question to present Aristotle as a forerunner of the above-mentioned authors (Craig, Beth, A. Robinson, R. C. Lyndon), but a renewed investigation of the text might reveal some striking logical intuitions.[23]

Dept. of Philosophy,
University of Salzburg and
Institut für Wissenschaftstheorie,
Internationales Forschungszentrum,
Salzburg, Austria

NOTES

* This is a revised version of a paper which I originally read at the Department of Philosophy at Stanford University in February 1978. At that time (from January to June 1978) I was invited as a guest-professor by the chairman (it was Joe Lambert) of the Department of Philosophy at the University of California at Irvine. I was then the 3rd exchange candidate of our exchange program between the Department of Philosophy at Irvine and ours at the University of Salzburg which we (Joe and me) founded in 1973 and which started in 1975. Since that time many scholars of both departments benefitted a lot from that exchange program. I want therefore to take the opportunity to say my warmest thanks to my friend Joe for the many troubles and for his permanent care for this important program. It is a great pleasure to me to dedicate an essay to him which I have prepared during my stay in Irvine.

1 Cf. K. Lambert, (MPI).

2 Cf. Aristotle (APo) II, ch. 2, 89b 36ss. Cf. the Introduction (by W. D. Ross), p. 609f. and 76.

3 Aristotle (APo) II, ch. 3. Cf. K. Lambert (MPI).

4 Aristotle (APo) II, ch. 4 and 6. Cf. W. D. Ross (APo), p. 629f.

5 Cf. Aristotle (APo) II, ch. 5.

6 (APo) II, 7, 92a 34—92b 3.

7 Ibid., 92b 3—92b 38.

8 There is one exception according to himself: God. As he says in Met 12, 6, 1071b

19, for the explanation of the moving universe a first principle is required whose essence is nothing but actual being (existence).

[9] Cf. (APo) II, 7, 92b 11—18. Cf. W. D. Ross (APo), p. 627. Cf. the reference to K. Lambert in note 3.

[10] Aristotle (APo) II, 7, 92b 19—25. Cf. the Commentary of Thomas Aquinas (AAC), Book II, lectures 6 and 7.

[11] This does not hold for all definitions. It holds not for nominal definitions or definitions as abbreviations which are not proper definitions according to him. Cf. (APo) II, 7, 92b 26—32. For modern views on definition see my (DRV).

[12] Cf. Aristotle (APo) II, 8, 93a 5—13. Porphyrius (Isg), ch. 2.

[13] Cf. note 12. See chapter 3,825 of my book (WTh) II, 1, and my (DRV). It is worth mentioning that those definitions which are used with great success in the sciences — since 2300 years if we remember the definition of the circle — seem much closer to the Aristotelean description than to some modern one, say that in *Principia Mathematica*, Vol. I, p. 11. There it is said, among other things, that definitions are neither true nor false but a matter of expediency. That this is even not correct in mathematics is shown by Kreisel in his (BMD). The whole history of the sciences shows that all the important and interesting definitions in science are true or false and therefore also subject to critical revision in the course of development of that science. Uninteresting definitions however like convenient abbreviations (say 'PTC-invariance' for 'parity-charge-time invariance') are not true or false. But it would seem to be much ado about nothing to construct a theory of definition for the latter.

[14] This was shown by V. F. Rickey (CDP), p. 276f.

[15] Cf. Aristotle (APo) II, ch. 8.

[16] (APo) II, ch. 1 and 2. Especially 90a 14—23, 90a 31—34 and 93a 3—4. Cf. W. D. Ross (APo), p. 611f.

[17] (APo) I, 14, 79a 24—29.

[18] (APo) I, 14, 79a 31 and 93a 3—8.

[19] Aristotle (APo) II, 8. Cf. W. D. Ross (APo), p. 77.

[20] For this and the next paragraph (8) cf. Thomas Aquinas (AAC) II, lecture 7, p. 187ff. The commentary of Ross to chapter 8 (93a 1—b 21) cf. (APo), p. 629ff. is not always clear.

[21] W. D. Ross (APo), p. 77.

[22] Craig (TUH),

[23] Beth (FMt), p. 293.

REFERENCES

Aristotle: 1949/1957, *Analytica Posteriora*, in: *Prior and Posterior Analytics*. intr. comm. W. D. Ross, Oxford. (APo)

Aristotle: 1958, *Metaphysics*, intr. comm. W. D. Ross, Oxford. (Met)

Beth, E. W.: 1968, *The Foundations of Mathematics*. Amsterdam. (FMt)

Craig, W.: 1957, 'Three Uses of the Herbrandt-Gentzen Theorem in Relating Model-Theory and Proof-Theory', *Journal of Symbolic Logic* **22**, 269—285. (TUH)

Kreisel, G.: 1981, 'Zur Bewertung mathematischer Definitionen', in: E. Morscher, O.

Neumaier, G. Zecha (eds.): *Essays in Science and Philosophy, dedicated to P. Weingartner*, Bad Reichenhall, pp. 185—209. (BMD)

Lambert, K.: 1983; *Meinong and the Principle of Independence*. Cambridge. (MPI)

Porphyrius: 1958, *Isagoge* (Introduction to the Categories), in: Aristoteles: *Kategorien, Lehre vom Satz* (ed., trad. E. Rolfes), Hamburg (Leipzig 1925), pp. 11—34. (Isg)

Rickey, V. F.: 1975, 'Creative Definitions in Propositional Calculi', *Notre Dame Journal of Formal Logic* **16**, pp. 273—294. (CDP)

Ross, W. D.: 1949/1957, *Prior and Posterior Analytics*, intr. comm., Oxford. (APo)

Thomas Aquinas: 1964, 'In Aristotelis Libros Posteriorum Analyticorum Expositio', in: *In Aristotelis Libros Petri Hermeneias et Posteriorum Analyticorum Expositio*, ed. R. M. Spiazzi, Marietti-Roma, Engl. (Transl. by F. R. Larcher, Magic Books, New York 1970). (AAC)

Weingartner, P.: 1989, 'Definitions in Russell, in the Vienna Circle and in the Lvov-Warsaw School', in: K. Szaniawski (ed.): *The Vienna Circle and the Lvov-Warsaw School*, Dordrecht, pp. 225—247. (DRV)

Weingartner, P.: 1976, *Wissenschaftstheorie* II, 1. Grundlagenprobleme der Logik und Mathematik. Stuttgart-Bad Cannstatt. (WTh) II.1

PETER W. WOODRUFF

ACTUALISM, FREE LOGIC AND FIRST-ORDER SUPERVALUATIONS

1. INTRODUCTION

In [Woodruff 1984] I showed that the supervaluation semantics introduced by [Van Fraassen 1966a, b] for Lambert's system of free logic [Meyer and Lambert 1968] failed to have certain metatheoretic properties (compactness, upward and downward Löwenheim-Skolem, strong completeness and recursive axiomatizability) which attach to ordinary first-order logic. I also introduced the idea of secondary supervaluations, in which the class of conventions for asigning truth values to formulas with non-designating terms was restricted in some way or another. I was able to show that for a particular such restriction, the falsity condition ($R(cx)$ false of all x when c doesn't denote), the desirable properties were regained. I speculated that the indifference condition ($R(cx)$ true either of all or of no x when c doesn't denote) would also suffice, and might be necessary.[1]

In this note I shall present a much more general sufficient condition, which I call actualist reduction. Along the way, I make some further refinements in the notion of convention introduced in [Woodruff 1984]. In essence, actualist reductionism allows only conventions which make statements involving non-denoting terms equivalent to some actualist statement. An informal example of such a convention would be a metalinguistic one, which decided the truth values of statements involving non-denoting terms in terms of some property (for example, occurring in a given story) of the statements themselves (which are, of course, existent objects). But the treatment here will be much more abstract than this particular example would indicate.

I do not believe that actualist reduction is a necessary condition for first-orderness, but will not present proof of that here. In the final section I shall indicate the direction in which I believe the next steps must be taken.

Wolfgang Spohn et al. (eds.), Existence and Explanation, 219—231.

2. SYNTAX: ACTUALIST FORMULAS, SKELETONS AND FICTIONAL ONTOLOGIES

We will be discussing a semantical treatment of a language $\mathscr{L}$ for first-order logic with identity and individual constants (but no other function symbols).[2] The existence predicate *E!* is defined contextually: *E!t* is $(\exists x)x \equiv t$, where x is the first variable not occurring in t. Among the formulas we distinguish the explicitly actualist formulas as follows. An occurrence of a constant c in φ is *guarded* if it is in a part of φ of the form $E!c$ or $E!c \rightarrow \psi$ or $E!c \wedge \psi$. A formula is explicitly actualist if every occurrence of a constant in it is guarded. As we will see, the truth value of such formulas in a model is entirely determined by what the model says about existents.

A *skeleton* is an atomic formula whose variables are, for some n, the first n variables of the language (in alphabetic order) occurring once each in left to right order; n is called the *degree* of the skeleton. A skeleton is *proper* if at least one constant occurs in it. We shall be interested in proper skeletons whose constants are non-denoting; such an one may be thought of intuitively as the predicate of existents determined by plugging up the relation of the skeleton with its non-existents in the way it does. σ is a skeleton *of* the atomic formula ψ iff the latter is the result of substituting some terms for the variables of σ; we also say ψ *has* the skeleton σ, or is an *instance of* σ. An occurrence of an instance of σ is guarded iff any of its constants is guarded.

The results below require that we make explicit the identities and non-identities holding between non-existents. To this end, we introduce the notion of a *fictional ontology*. This consists of a set N of individual constants, together with an equivalence relation I on N. Intuitively, N is a set of non-denoting constants, thought of as if they were names of non-existent objects, and I specifies which names are names of 'the same' non-existent object. The ontology (N, I) is said to be *over* N.

An ontology is *for* a skeleton if it includes all the constants of the skeleton. For any atomic formula $\varphi = Ft_1 \ldots t_n$ and ontology (N, I) the skeleton of φ obtained by replacing just those terms t_i which do not occur in N by distinct variables will be called the skeleton of φ in that ontology. A *minimal* ontology for a skeleton σ is one in which every member of N is equivalent to some constant occurring in σ. The *trace* of a skeleton in an ontology for it is simply the restriction of that ontology to the constants equivalent to those of the skeleton; evidently traces are always minimal.

A *diagram* of a fictional ontology (N, I) consists of the non-existence claims $\neg E!c$ for c in N, together with the identities $c \equiv d$ such that cId and the non-identities $\neg c \equiv d$ between the remaining pairs in N. We will call a set of formulas a diagram if it is the diagram of any ontology; it is clear that a diagram is a diagram of exactly one ontology. Thus all the previous terminology can be applied to diagrams as well. If a diagram is finite, we will not usually distinguish the diagram from the conjunction of its members. Thus a minimal diagram may always be regarded as a formula.

3. SEMANTICS: PARTIAL MODELS, CONVENTIONS AND CLASSICAL SATISFACTION

A partial model $\mathfrak{A}$ is a pair $\langle A, v \rangle$, A a set and v an interpretation which assigns to each predicate an extension (of appropriate degree) in A, and to some (perhaps all or no) constants references in A. A *total* model is one which assigns every constant a reference. An assignment $\mathbf{a}$ in A maps each variable to a member of A. The language of $\mathfrak{A}$, $\mathscr{L}_{\mathfrak{A}}$, consists of those formulas all of whose constants are denoting in $\mathfrak{A}$. We shall write $N_{\mathfrak{A}}$ for the set of non-referring constants of $\mathfrak{A}$.

A partial model is of course an ordinary classical model for its language (as just defined); the problem is to assign interpretations to expressions not in this language, i.e., expressions containing non-denoting terms. The salient idea of super-valuation semantics since [Van Fraassen 1966a] has always been to treat non-denoting singular terms contextually, assigning interpretations (by convention) only to formulas containing such terms, rather than to the terms themselves. This was carried out in an ingenious but somewhat imperspicuous manner in [Van Fraassen 1966b]; I refined the method in [Woodruff 1984] and now propose a further refinement.

The idea is to divide conventions into two parts. First we make certain conventions about which non-denoters are to be treated 'as if' co-referential; this takes the form of a fictional ontology over $N_{\mathfrak{A}}$. Then we treat skeletons of this ontology (which we call $\mathfrak{A}$-skeletons) as if they were predicates added to $\mathscr{L}_{\mathfrak{A}}$.[3] In so doing, we must make sure that predicates involving 'co-referential' non-denoters are co-extensional. This leads to the following definitions.

A fictional ontology over the nondenoting constants of a model is also said to be over the model. A *convention* C for $\mathfrak{A}$ has two parts: a fictional ontology over $\mathfrak{A}$ (we say the ontology is *covered* by C) and an interpretation (also designated C) which assigns extensions (of appropriate degree) to skeletons of this ontology, subject to the following conditions:

a) $C(c \equiv x) = C(x \equiv d) = \emptyset$

b) $C(c \equiv d) = 1$ iff c is equivalent to d in the ontology of C

c) if σ is equivalent to σ' in the ontology of C, then $C(\sigma) = C(\sigma')$

Let $\mathfrak{A}$ be a partial structure and C a convention over $\mathfrak{A}$. Then we define the classical satisfaction relation for $\mathfrak{A}$ and C as follows. If φ is atomic, then $(\mathfrak{A}, C) \models \varphi[\mathbf{a}]$ iff either φ is in $\mathscr{L}_{\mathfrak{A}}$ and $\mathfrak{A} \models \varphi[\mathbf{a}]$ in the usual sense, or $\varphi = \sigma(t_1, \ldots, t_n/x_1, \ldots, x_n)$, for σ an $\mathfrak{A}$ skeleton and all t_i in $\mathscr{L}_{\mathfrak{A}}$, and $\langle t_1[\mathbf{a}], \ldots t_n[\mathbf{a}]\rangle$ belongs to $C(\sigma)$. For complex φ we then have the usual recursive definition.

The following lemmas emphasize the analogy between the skeletons and second-order variables.

LEMMA 3.1. Let C and C' assign the same extensions to skeletons with instances in φ. Then $(\mathfrak{A}, C') \models \varphi[\mathbf{a}]$ iff $(\mathfrak{A}, C) \models \varphi'[\mathbf{a}]$.

Proof. by induction on φ, is straightforward. q.e.d.

LEMMA 3.2. Let σ be an $\mathfrak{A}$-skeleton, ψ any formula, C any convention, and C' like C save that it assigns to σ the extension ψ has on C and $\mathbf{a}$. That is, if σ is n-ary, $C'(\sigma) = \{\langle \mathrm{d}_1, \ldots \mathrm{d}_n\rangle: (\mathfrak{A}, C) \models \psi[\mathbf{a}(\mathrm{d}_1, \ldots, \mathrm{d}_n/x_1, \ldots, x_n)]\}$. Then for any formula φ, $(\mathfrak{A}, C') \models \varphi[\mathbf{a}]$ iff $(\mathfrak{A}, C') \models \varphi'[\mathbf{a}]$, where φ' is the result of replacing every instance of σ in φ by a corresponding instance of ψ (changing bound variables in ψ as incessary to avoid capture of free variables).

Proof. by induction on φ. If σ doesn't occur in φ, then $\varphi = \varphi'$ and the result follows from 3.1. Otherwise, if φ is atomic, it must itself be an instance of σ, say $\sigma(t_1, \ldots, t_n/x_1, \ldots, x_n)$. Then φ' is (up to changes of bound variable) $\psi(t_1, \ldots, t_n/x_1, \ldots, x_n)$. Now $(\mathfrak{A}, C') \models \varphi[\mathbf{a}]$ iff $\langle t_1[\mathbf{a}], \ldots, t_n[\mathbf{a}]\rangle$ belongs to $C'(\sigma)$, iff (by def of $C\boldsymbol{\varphi}$) $(\mathfrak{A}, C) \models \psi[\mathbf{a}(t_1[\mathbf{a}], \ldots, t_n[\mathbf{a}]/x_1, \ldots, x_n)]$, iff (by the usual lemma on substitution) $(\mathfrak{A}, C) \models \varphi'[a]$. The rest of the induction is straightforward. q.e.d.

We may also verify the claim about actualist formulas made above; in fact, something slightly stronger. Let φ be actualist in $\mathfrak{A}$ if every occurrence of an $\mathfrak{A}$-non-denoting constant is guarded. And let φ be *stable* in $\mathfrak{A}$ if for all C, C' over $\mathfrak{A}$, and assignments **a**, $(\mathfrak{A}, C) \models \varphi[\mathbf{a}]$ iff $(\mathfrak{A}, C') \models \varphi[\mathbf{a}]$.[4]

LEMMA 3.3. If φ is actualist in $\mathfrak{A}$, then it is stable in $\mathfrak{A}$.

Proof. by induction on φ. If the latter is atomic, then all occurrences of constants in it are denoting, so the conclusion follows by definition of $\models$ above. If φ is $E!c \wedge \psi$,[5] then if c is defined in $\mathfrak{A}$, ψ is actualist in $\mathfrak{A}$ and the conclusion follows by inductive hypothesis and the fact that when c is defined, $(\mathfrak{A}, C) \models (E!c \wedge \psi)[\mathbf{a}]$ iff $(\mathfrak{A}, C) \models \psi[\mathbf{a}]$; but if c is undefined, then not $(\mathfrak{A}, C) \models (E!c \wedge \psi)[\mathbf{a}]$ for any C. Otherwise, if φ is $\neg\psi$, then ψ must be actualist, so the conclusion follows from the inductive hypothesis. If φ is $E!c \rightarrow \psi$, then the argument is similar to that just given, while otherwise, if φ is $\psi \vee \theta$, both disjuncts must be actualist in $\mathfrak{A}$ and the conclusion follows from inductive hypothesis. Finally, if φ is $E!c$, then it is stable by condition 3b) on conventions; otherwise, if φ is $(\exists x)\psi$, ψ must be actualist and we use inductive hypothesis. q.e.d.

It is an immediate corollary to the lemma that all explicitly actualist wffs are everywhere stable.

4. CONVENTIONAL POLICIES AND SUPERSATISFACTION

If $\mathscr{C}$ is a class of conventions over $\mathfrak{A}$, we say $(\mathfrak{A}, \mathscr{C}) \models \varphi[\mathbf{a}]$ iff for all C in ψ, $(\mathfrak{A}, C) \models \varphi[\mathbf{a}]$. A *conventional policy* $\mathscr{P}$ is a function associating with each structure $\mathfrak{A}$ some set of conventions over $\mathfrak{A}$.[6] With each policy are associated two relations of logical consequence and a satisfaction relation, given by the following definitions:

$$\Sigma \models_{\mathscr{P},c} \varphi \quad \text{iff for all } \mathfrak{A}, \mathbf{a}, \text{ and } C \text{ in } \mathscr{P}(\mathfrak{A}), (\mathfrak{A}, C) \models \Sigma[\mathbf{a}] \text{ only if } (\mathfrak{A}, C) \models \varphi[\mathbf{a}]$$

$$\mathfrak{A} \models_{\mathscr{P}} \varphi[\mathbf{a}] \text{ iff } (\mathfrak{A}, \mathscr{P}(\mathfrak{A})) \models \varphi[\mathbf{a}]$$

$$\Sigma \models_{\mathscr{P},s} \varphi \quad \text{iff for all } \mathfrak{A}, \mathbf{a}, \mathfrak{A} \models_{\mathscr{P}} \Sigma[\mathbf{a}] \text{ only if } \mathfrak{A} \models_{\mathscr{P}} \varphi[\mathbf{a}]$$

The first notion is called classical consequence on the policy $\mathscr{P}$, the second super-satisfaction and the third super-consequence on $\mathscr{P}$.

If $\mathfrak{A}$ is total, then $\mathscr{P}(\mathfrak{A})$ must obviously be empty. If that is the only case in which $\mathscr{P}(\mathfrak{A})$ is empty, I call $\mathscr{P}$ strongly consistent. A strongly consistent policy always allows some conventional arrangements about non-denoting terms, whatever is the case in the 'real world'. A more subtle property of policies is *identity neutrality*, where a policy is identity neutral at $\mathfrak{A}$ if every $\mathfrak{A}$-ontology is covered by some convention, and a policy is identity-neutral (*simpliciter*) if it is identity-neutral at every model. Evidently, every identity neutral policy is strongly consistent.

Consistency and identity neutrality are examples of properties of a policy which would be relevant to its acceptability for purposes of logical or philosophical analysis. The identification and investigation of such properties seems to me to be the most important direction for future research in supervaluation theory. But in the present work we shall consider only one more property: definability.

5. DEFINABLE POLICIES

A policy $\mathscr{P}$ is (*first-order*) *definable* if it may be characterized as follows:

$$\mathscr{P}(\mathfrak{A}) = \{C : (\mathfrak{A}, C) \models \Delta\}$$

where Δ is some set of sentences, called the *constraints* imposed by the policy. Some examples:

The trivial policy $\mathscr{T}: \Delta = \emptyset$

The falsity policy $\mathscr{F}$: $\Delta = \{\neg E!c \rightarrow (\forall x_1, \ldots, x_k)\neg \varphi : h$ a non-logical atomic formula in which both c and $x_1, \ldots, x_k$ occur, and containing only those variables$\}$

The verity policy $\mathscr{V}: \Delta = \{\neg E!c \rightarrow (\forall x_1, \ldots, x_k)\varphi :$ same proviso$\}$

The all-or-none policy $\mathscr{AN}: \Delta = \{\neg E!c \rightarrow (\forall x_1, \ldots, x_k)\varphi \vee (\forall x_1, \ldots, x_k)\neg \varphi :$ same proviso$\}$

These policies are straightforward but rather procrustean. We now describe a family of policies which implicitly includes those above but allows a great more flexibility. Call a pair (δ, σ) *apt* if δ is a minimal diagram for the skeleton σ and σ is not an identity. Let f be an effective map from apt pairs to finite sequences of actualist formulas whose free variables are among those of σ. Further, let f have the same value for δ-equivalent skeletons. Then there is an *actualist reduction* policy over f, the f-policy, for short, whose constraints are all sentences of the following form:

$$\delta \rightarrow \cdot (\forall x_1, \ldots, x_k)(\sigma \leftrightarrow \psi_1) \vee \ldots \vee (\forall x_1, \ldots, x_k)(\sigma \leftrightarrow \psi_m)$$

where $\psi_1, \ldots, \psi_m$ is $f(\delta, \sigma)$ and the displayed variables are those occurring in σ. Intuitively, $f(\delta, \sigma)$ gives the conventional options open to us in interpreting σ for the ontology δ; it is actualist because these options are given by actualist formulas, reductionist because they are given as equivalences.

We observe that the verity, falsity and all-or-none policies are (almost) actualist reduction policies, since we can write $(\forall x) \neg \varphi$ as $(\forall x).\ \varphi \leftrightarrow \bot$, and $(\forall x)\varphi$ as $(\forall x).\ \varphi \rightarrow \top$, where $\top$ is defined as $(\forall x)\ x \equiv x$ and $\bot$ as $\neg\top$. The trivial policy yields the ordinary notions of classical and super-consequence for free logic, and we will omit the subscript $\mathcal{T}$ from the corresponding consequence and satisfaction relations.

The above policies are uniform or symmetric in that they treat all constants alike. There is some reason to suppose that anything that really counts as a 'policy' would have this kind of uniformity. But in any case, our notion also includes very different kinds of policies; for instance those in which the constraints are the text of some story, theory or myth.

The first of the following lemmas records (without proof) some simple facts about the consequence relations associated with definable policies; the second demonstrates an important property of the particular policies given above which will be essential to our main result.

LEMMA 5.1. Let $\mathcal{P}$ have constraints Δ. Then

a) $\Sigma \models_{\mathcal{P},c} \varphi$ iff $\Sigma, \Delta \models_c \varphi$

b) if $\Sigma \models_{\mathcal{P},c} \varphi$, then $\Sigma \models_{\mathcal{P},s} \varphi$

c) $\models_{\mathcal{P},c} \varphi$ iff $\models_{\mathcal{P},s} \varphi$ iff $\Delta \models_c \varphi$

d) if $\Delta \subseteq \Delta'$, and $\mathcal{Q}$ has constraints Δ', then $\mathfrak{A} \models_{\mathcal{P}} \varphi[\mathbf{a}]$ only if $\mathfrak{A} \models_{\mathcal{Q}} \varphi[\mathbf{a}]$

e) If $\Delta \models_c \varphi \leftrightarrow \psi$, then $\mathfrak{A} \models_{\mathcal{P}} \varphi[\mathbf{a}]$ iff $\mathfrak{A} \models_{\mathcal{P}} \psi[\mathbf{a}]$

LEMMA 5.2. Every actualist reduction policy is identity neutral.

Proof. Let a partial model $\mathfrak{A}$ and an f-policy be given, together with an $\mathfrak{A}$-ontology. We define a convention C as follows: the ontology of C is the given one; conditions a) and b) are to be satisfied (as they clearly can be). Otherwise, let σ be an $\mathfrak{A}$-skeleton, and τ its trace in the given ontology. Then (τ, σ) is an apt pair, and we set $C(\sigma) = \{\langle d_1, \ldots d_n\rangle : \mathfrak{A}$

$\models_s \psi[\mathbf{a}(d_1, \ldots, d_n/x_1, \ldots, x_n)]\}$, where **a** is any assignment and ψ is the first formula in $f(\tau, \sigma)$. Conditions a) and b) on conventions are satisfied by construction; and c) is satisfied by construction and the fact that when σ is equivalent to σ', they have the same trace. Thus C is in fact a convention. To see that the constraints of the policy are satisfied, consider

$$\gamma \rightarrow \cdot (\forall x_1, \ldots x_k)(\sigma \leftrightarrow \psi_1) \vee \ldots \vee (\forall x_1, \ldots, x_k)(\sigma \leftrightarrow \psi_m)$$

If γ is the trace of σ, the first disjunct of the consequent will be satisfied by C. For if $(\mathfrak{A}, C) \models \sigma[\mathbf{a}]$, then by definition of C, $\mathfrak{A} \models_s \psi[\mathbf{a}]$, and so since C is a convention, $(\mathfrak{A}, C) \models \psi[\mathbf{a}]$ (by definition of supersatisfaction.) But if not, then by similar reasoning, not $\mathfrak{A} \models_s \psi[\mathbf{a}]$. Now ψ is, as a member of $f(\gamma, \sigma)$, an actualist formula and thus, by lemma 3.3, stable. Hence for all C', not $(\mathfrak{A}, C') \models \psi[\mathbf{a}]$. But then in particular, not $(\mathfrak{A}, C) \models \psi[\mathbf{a}]$.

If γ is not the trace of σ in the ontology of C, then it will differ from that ontology in some identity, and hence will be falsified by C. Thus in either case, the constraint is satisfied. q.e.d.

6. A NORMAL FORM

In this section we will show that every formula has an equivalent normal form in which alternative ontologies are made explicit. We proceed as follows: given a formula, we first conditionalize it by all possible ways in which its unguarded contants might denote or not. Next, we separate each of these conditions into existence and non-existence claims, and export the latter. Finally, we conditionalize further on the ways in whcih non-existents could be identical or not.

LEMMA 6.1. a) Let a prefix for φ be a conjunction of existence and non-existence claims for the unguarded constants of φ, one for each. Then the disjunction of all prefixes for φ is a tautology, and φ is tautologically equivalent to the conjunction of all formulas $\pi \rightarrow \varphi$, π a prefix for φ.

b) Let π_1 contain the non-existence claims, and π_2 the existence claims in π. Then $\pi \rightarrow \varphi$ is equivalent to $\pi_1 \rightarrow . \pi_2 \rightarrow \varphi$.

Proof. by truth-functional logic.

LEMMA 6.2. Let N be a finite set of constants, ν its non-existence claim.[7] Then

a) ν is classically equivalent to the disjunction of all diagrams over N.

b) $\nu \to \varphi$ is classically equivalent to the conjunction of all $\delta \to \varphi$, δ a diagram over N.

Proof: a) let a proto-diagram over N be a conjunction of ν with one of $c \equiv d$ or $\neg c \equiv d$ for each pair c, d of constants in N. Then the disjunction of all proto-diagrams over N is tautologically equivalent to ν. But a proto-diagram which fails to be a diagram (because it violates symmetry, reflexivity or transitivity) will be false on every convention, either because ν is false or by condition b) on conventions. Thus the disjunction of all proto-diagrams is equivalent to the disjunction of all diagrams. b) now follows by truth-functional logic. q.e.d.

We call a formula *explicit* if it is of the form $\delta \to \varphi$, where δ is a diagram for the unguarded constants in φ. And we say that a formula is in *explicit normal form* if it is a conjunction of explicit formulas. Then the above lemmas have the following corollary:

Theorem 1. Every formula is classically equivalent to a formula in explicit normal form

7. ACTUALIST POLICIES ARE STABLE

We now show that all actualist-reduction policies are stable, in the sense that for any formula φ, there will be a stable formula super-equivalent to it on that policy. From this the first-order character of actualist semantics will follow.

Theorem 1 tells us that any formula is classically (and hence, by 5.1 e), super-equivalent to a conjunction of explicit formulas. Furthermore, conjunctions of super-equivalent formulas are super-equivalent. It will thus suffice to show that every explicit formula is super-equivalent to a stable formula. Let $\delta \to \varphi$ be explicit. We show that $\delta \to \varphi$ is super-equivalent to a conjunction of formulas $\delta \to \psi$ in which ψ is actualist, and hence stable. The proof is by induction on the number of δ-skeletons which have unguarded instances in φ. If there are none, then there can be no unguarded instances of constants in φ. So let σ be such a skeleton.

If σ is an identity, then we note that $\delta \to .\ \forall x(c \equiv x \leftrightarrow \bot)$, $\delta \to .\ \forall x(x \equiv c \leftrightarrow \bot)$, $\delta \to .\ c \equiv d \leftrightarrow \top$, and $\delta \to .\ c \equiv d \leftrightarrow \bot$ are classically valid, where c is a δ-nonexistent and $c \equiv d$ is respectively an identity or non-identity in δ. Thus $\delta \rightarrow \varphi$ will be equivalent to $\delta \rightarrow \varphi'$, where φ', where φ' is the result of replacing occurrences of $x = c$ by $\perp$ and so on. But φ' will contain fewer unguarded skeletons, and so we may apply the inductive hypothesis.

Otherwise, let τ be the trace of σ in δ. Then (τ, σ) is an apt pair, hence there is an f-constraint

$$\tau \rightarrow \cdot (\forall x_1, \ldots, x_k)(\sigma \leftrightarrow \psi_1) \vee \ldots \vee (\forall x_1, \ldots, x_k)(\sigma \leftrightarrow \psi_{\mathrm{m}})$$

which all conventions must satisfy. Let φ_i be the result of replacing σ by ψ_i in φ. I claim that $\delta \rightarrow \varphi$ is super-equivalent to the conjunction of the formulas $\delta \rightarrow \varphi_i$, $i \leqslant m$.[8]

Suppose first $\mathfrak{A} \models_f \delta \rightarrow \varphi[\mathbf{a}]$, C is any f-convention over $\mathfrak{A}$, and $j \leqslant m$. Then if $(\mathfrak{A}, C) \models \delta[\mathbf{a}]$, let C' be like C except for assigning to σ the extension ψ_j has in C. If C is an f-convention, C' must be so as well. First of all, C and C' have the same ontology, hence satisfy the same diagrams. Furthermore, since the reducing formulas of any constraint are stable (by definition), by lemma 3.3 every convention satisfies (on every assignment) the same such formulas as C. Thus, since C' does not change the extension of any skeleton but σ, and C is an f-convention, C' will satisfy all non-σ constraints. But a convention can satisfy at most one minimal diagram for a given skeleton. So if η is any minimal σ-diagram other than τ, then C' falsifies η and hence satisfies the η, σ constraint. Finally, C' satisfies the τ, σ constraint given above because (by construction and lemma 3.3) it satisfies $(\forall x_1, \ldots, x_k)(\sigma \leftrightarrow \psi_j)$. Now by hypothesis $\mathfrak{A} \models_f \delta \rightarrow \varphi[\mathbf{a}]$, so we must have $(\mathscr{A}, C') \models \delta \rightarrow \varphi[\mathbf{a}]$; and also by hypothesis and the above, C' satisfies δ, so we have $(\mathfrak{A}, C') \models \varphi[\mathbf{a}]$. But then by lemma 3.1, $(\mathfrak{A}, C) \models \varphi_j[\mathbf{a}]$. So $\mathfrak{A} \models_f \delta \rightarrow \varphi_j[\mathbf{a}]$.

Conversely, if not $\mathfrak{A} \models_f \delta \rightarrow \varphi[\mathbf{a}]$, then there is some C which satisfies all f-constraints and δ, but not φ. Since it satisfies δ, it must satisfy τ, and hence also $(\forall x_1, \ldots, x_k)(\delta \leftrightarrow \psi_j)$, for some $j \leqslant m$. But then C itself assigns the same extension to σ as to ψ_j, and so by lemma 3.1, C must fail to satisfy φ_j and hence also $\delta \rightarrow \varphi_j$.

Now since ψ_j is actualist, each φ_j has one fewer unguarded skeletons than φ. Thus we can appy the inductive hypotesis.

Suppose finally that φ is stable. Then I claim that $\delta \rightarrow \varphi$ is super-equivalent to φ. Since the f-policies are identity neutral by lemma 5.2, every diagram is satisfied by some convention. If $\delta \rightarrow \varphi$ is super-satisfied in a model, then some convention satisfies δ and hence φ;

since φ is stable, it must also be super-satisfied in that model. And evidently, $\delta \rightarrow \varphi$ is super-satisfied if φ is.

Putting all the above together, we see that our original formula φ turns out to be f-equivalent to the conjunction of the results of replacing its unguarded skeletonsin all possible ways according to the f-constraints.[9] Since these conjuncts are all actualist (and hence, stable), we have:

Theorem 2. Let $\mathscr{P}$ be an actualist reduction policy. Then every formula of free logic is $\mathscr{P}$-equivalent to an actualist (and hence, stable) formula.

8. CONCLUSION

Theorem 2 at once gives the compactness and Löwenheim-Skolem theorems for actualist reduction semantics. For the first, we have the argument of [Bencivenga 1983], Lemma 1. For the second, suppose $\mathfrak{A} \models_f \varphi$ and h is stable. Then since f-policies are strongly consistent, there is a convention C for which $(\mathfrak{A}, C) \models \varphi$. By theorem 6 of [Van Fraassen 1968], there is a denumerable model $\mathfrak{A}'$ and a convention C' over $\mathfrak{A}'$ which together satisfy φ. But then by the stability of φ, it is super-satisfied in $\mathfrak{A}'$.

The argument just given uses only the stability of φ; thus any policy on which every formula is super-equivalent to a stable formula will give the same results. A likely conjecture is that the converse holds as well: that if compactness and L-S hold for a given policy, then every formula is equivalent on that policy to a stable one. A further likely conjecture is that the stable formulas coincide with those which are classically equivalent to an explicitly actualist formula.

The next step in super-valuation theory for free logic is clearly to settle these conjectures. The first is related to Lindström's Theorem [Lindström 1969], which shows that among the various (classical) logics (infinitary, second-order, generalized quantifier, etc.) which contain first-order logic, the latter is the only one to sport both the compactness and L-S theorems. However, the theorem is not directly applicable in the present case, since we have different semantics for the same language, rather than different languages for (in effect) the same semantics. Nevertheless, I believe that Lindström's methods can be generalized to provide proof of the first conjecture.[10]

The second conjecture is related to the result of [Dwyer 1989] that a sentence of free logic is significant (in his sense) iff it is equivalent to a basic sentence (in his sense). It can be shown that every basic sentence is equivalent to an explicitly actualist one, and vice versa; furthermore, significance and our notion of stability are closely related. It is therefore likely that Dwyer's results can be adapted to the present case.

Finally, let me repeat the remark of 5, that from a philosophical point of view the most interesting direction for research is in the investigation of properties of conventional policies, andin the proposing of particular conventions for applied logics. Only in this way will one be able to assess whether the *prima facie* attractions of conventionalism and actualism which, super-valuations seem to provide, do in fact stand up under close scrutiny.

Department of Philosophy
University of California,
Irvine, U.S.A.

NOTES

[1] In the present note I call this the 'all-or-none' condition. I would like to note here that the argument of this paper was greatly influenced by a reading of [Bencivenga 1983]; in particular the emphasis on stability, and the strategy of proviing compactness by showing equivalence to stable formulas is due to him. In [Woodruff 1984] I claimed that Bencivenga's results were obtainable from the normal form theorem of that paper, but this claim was incorrect; something like the argument of the present paper is required to get from a 'first-order' normal form to compactness. In general, it seems that the normal form theorem is convenient for showing supervaluational semantics to be second-order; when we wish to show that certain restricted forms of the semantics are still first-order, quite other methods are required.

[2] For purposes of arguments by syntactical induction, we will assume that $\neg$, v, $\exists$ and $\equiv$ are the primitive logical symbols, the last standing for identity.

[3] This is possible because every atomic formula of $\mathscr{L}_{\mathfrak{A}}$ is uniquely of the form $\sigma(t_1, \ldots t_n/x_1, \ldots, x_n)$ for some $\mathfrak{A}$-skeleton σ, where all the t_i belong to $\mathscr{L}_{\mathfrak{A}}$.

[4] See Note 1.

[5] I.e., $\neg(\neg E!c \vee \neg\psi)$.

[6] This is not the sense of 'policy' used in Van Fraassen [1969]. Our policies provide what he calls "information about not-existents". (§4)

[7] I.e., the conjunction of the non-existence claims for its members.

[8] The point of this step may be seen by recalling that $\forall x(x = a \vee x = b \rightarrow Fx)$ is equivalent to $Fa \wedge Fb$. Skeletons are second-order variables; supervaluations universally quantify them, and the constraints have the effect of restricting that quantification to the particular relations expressed by the formulas ψ_i.

[9] In effect, we have given a kind of 'quantifier elimination', though the quantifiers here are implicit (in supervaluationsp and second-order, and the result of the elimination is first-orderness, rather than decidability. Cp. fn. 8.
[10] It is quite another question, even if the conjecture is true, whether we can give a syntactical criterion; i.e. whether every policy which allows stabilization is definable in some uniform way. I have no conjecture to make in this regard.

REFERENCES

Bencivenga, Ermanno: 1983, 'Compactness of a supervaluational language', *Journal of Symbolic Logic* **48**, 384—386.
Dwyer, Robert: 1989, *Denoting and Defining*, PhD Dissertation, University of California, Irvine.
Meyer, Robert and Lambert, Karel: 1968, 'Universally free logic and standard quantification theory', *Journal of Symbolic Logic* **33**, 8—26.
Lindström, Per: 1969, 'On extensions of elementary logic', *Theoria* **35**, 1—11.
Van Fraassen, Bas: 1969a, 'Singular terms, truth value gaps and free logic', *Journal of Philosophy* **63**, 481—495.
Van Fraassen, Bas: 1969b, 'The completeness of free logic', *Zeitschrift für Mathematische Logik und Grundlagen der Mathematik* **12**, 219—234.
Van Fraassen, Bas: 1960, 'A topological proof the Löwenheim-Skolem, compactness and strong completeness theorems for free logic', *ibid.* **14**, 245—254.
Van Fraassen, Bas: 1969, 'Presuppositions, supervaluations and free logic', *The Logical Way of Doing Things*, K. Lambert (ed.), Yale University Press, New Haven, Connecticut, pp. 67—91.
Woodruff, Peter: 1984, 'On supervaluations in free logic', *Journal of Symbolic Logic* **49**, 943—950.

BIBLIOGRAPHY OF KAREL LAMBERT

1. BOOKS

1969: *The Logical Way of Doing Things* (Editor), Yale University Press, New Haven.
1970: *Philosophical Problems in Logic* (Editor), Reidel, Dordrecht. Reprinted 1982.
1970: *An Introduction to the Philosophy of Science* (with G. Brittan), Prentice-Hall, Englewood Cliffs.
 a. *Portuguese Edition*, Cultrix, Sao Paulo, 1972.
 b. *Spanish Edition*, Guadarrama, Madrid, 1974.
 c. *Japanese Edition*, Baifu Kan, Tokyo, 1981.
1972: *Derivation and Counterexample* (with B. van Fraassen), Dickenson, Encino.
 a. *Chinese Edition*, Taiwan, 1978.
1979: *Introduction to the Philosophy of Science* (Second Edition) (with G. Brittan), Ridgeview, Atascadero, CA.
 a. *Italian Edition*, Boringhieri, Torino, 1979.
 b. *Japanese Edition*, Koyoshobo, Kyoto, 1982.
1980: *The Nature of Argument* (with W. Ulrich), MacMillan, New York.
1983: *Meinong and the Principle of Independence*, The University Press, Cambridge.
1987: *Logic, Bivalence and Denotation* (with E. Bencivenga and B. van Fraassen), Ridgeview Atascadero, CA.
1987: *Introduction to the Philosophy of Science* (Third Edition) (with G. Brittan), Ridgeview, Atascadero, CA.
 a. *German Edition*, de Gruyter, Berlin, 1991.
1991: *Philosophical Applications of Free Logic* (Editor), Oxford University Press, New York.

2. ARTICLES

1956: 'On Naming and Claiming', *Philosophical Studies* **7**, 43—46.
1956: 'Synonymity Again', *Analysis* **XVI**, 71–72.
1958: 'Notes on 'E!': I', *Philosophical Studies* **9**, 60—63.
1958: 'An Experimental Study of What is Learned in a Shuttlebox Situation' (with D. Gorfein), *Canadian Journal of Psychology* **12**, 222—228.
1959: 'Singular Terms and Truth', *Philosophical Studies* **10**, 1—5.
1960: 'A Study of Latent Inference Learning', *Canadian Journal of Psychology* **14**, 45—50.
1961: 'Notes on 'E!': II', *Philosophical Studies* **12**, 1—5.
1961: 'The Effect of Hunger Satiation in the Test Phase on Inference Performance' (with C. N. Uhl), *Canadian Journal of Psychology* **15**.
1962: 'Notes on 'E!': III', A Theory of Descriptions', *Philosophical Studies* **13**, 51—59. Reprinted in *Le Logiche Libere* (edited by Ermanno Bencivenga), Boringhieri, Torino (1976).

Wolfgang Spohn et al. (eds.), Existence and Explanation, 233—236.

1963: 'Existential Import Revisited', *Notre Dame Journal of Formal Logic* **4**, 288—292.
1963: 'Explaining Away Singular Non-existence Statements', *Dialogue* **1**, 381—389.
1963: 'Some Remarks on Singular Terms: A Critical Review of H.S. Leonard's *Principles of Right Reason*', *Inquiry* **6**, 257—272.
1963: 'Quantification and Existence', *Inquiry* **6**, 319—324.
1964: 'Notes on 'E!': IV', A Reduction in Free Quantification Theory With Identity and Definite Descriptions', *Philosophical Studies* **15**, 85—87. Reprinted in *Le Logiche Libere* (edited by Ermanno Bencivenga), Boringhieri, Torino (1976).
1965: 'Logic and Existence', *Notre Dame Journal of Formal Logic* **6**, 135—141
1965: 'Is Mathematics a Science?', *Proceedings of the West Virginia Academy of Science* **38**, 29—34.
1966: 'On the Noncommunicator', *Philosophical Studies* **17**, 27—30.
1966: 'Identity and Definite Descriptions II', *Philosophical Studies* **17**, 35—43.
1967: 'Free Logic and the Concept of Existence', *Notre Dame Journal of Formal Logic* **8**. Reprinted in *Le Logiche Libere* (edited by Ermanno Bencivenga), Boringhieri, Torino (1976).
1967: 'On Free Description Theory', (with B. van Fraassen), *Zeitschrift für Mathematische Logik und Grundlagen der Mathematik* **13**, 225—240. Reprinted in *Le Logiche Libere* (edited by Ermanno Bencivenga), Boringhieri, Torino (1976).
1968: 'A Translation Theorem for Two Systems of Free Logic' (with T. Scharle), *Logique et Analyse* **39—40**, 8—26. Reprinted in *Le Logiche Libere* (edited by Ermanno Bencivenga), Boringhieri, Torino (1976).
1968: 'Universally Free Logic and Standard Quantification Theory' (with R. Meyer), *Journal of Symbolic Logic* **33**, 8—26. Reprinted in *Le Logiche Libere* (edited by Ermanno Bencivenga) Boringhieri, Torino (1976).
1968: 'On the No-Type Theory of Significance', *Australasian Journal of Philosophy* **46**, 79—86.
1969: 'A Liberated Version of S5' (with H. Leblanc and R. Meyer), *Archiv für Mathematische Logik und Grundlagenforschung* **12**, 151—154. Reprinted in *Existence, Truth and Provability* (edited by H. Leblanc) State University of New York Press, Albany (1981).
1969: 'Logical Truth and Microphysics', in *The Logical Way of Doing Things* (edited by Karel Lambert), Yale University Press, New Haven.
1970: 'Meaning Relations, Possible Objects and Possible Worlds', (with B. van Fraassen in *Philosophical Problems in Logic* (edited by Karel Lambert), Reidel, Dordrecht.
1971: 'Explanation and Intention', in *The Psychology of Private Events* (edited by A. Jacobs and L. Sacks), Academic Press, New York.
1972: 'Being and Being So', in *Jenseits von Sein und Nichtsein* (edited by Rudolf Haller) Akademische Druck u. Verlagsanstalt, Graz.
1972: 'Notes on Free Description Theory', *Journal of Philosophical Logic* **1**, 184—191.
1973: 'The Theory of Objects: A Critical Review of J. N. Findlay's *Meinong's Theory of Objects and Values*', *Inquiry* **16**, 220—230.
1974: 'Predication and Extensionality', *Journal of Philosophical Logic* **3**, 255—264; reprinted in *Problems in Logic and Ontology* (edited by E. Morscher *et al.*), Akademische Druck u. Verlaganstalt, Graz.
1974: 'Impossible Objects', *Inquiry* **17**, 303—314.

1976: 'On 'The Durability of Impossible Objects' ', *Inquiry* **19**, 251—255.
1977: 'Unmögliche Gegenstande. Eine Untersuchung der Meinong-Russell-Kontroverse', in *Conceptus: Österreichische Philosophen und ihr Einfluß auf die analytische Philosophie der Gegenwart*, pp. 92—100.
1978: 'Review of Susan Haack's *Deviant Logic*', *Journal of Symbolic Logic* **43**, 377—379.
1978: 'The Place of Intention in the Explanation of Behavior', *Grazer Philosophische Studien* **6**, 75—85.
1979: 'The Status of Analytic Philosophy in Contemporary Austria', in *Austriaca-Americana* **5**, 135—139.
1980: 'Explanation and Understanding: An Open Question?', in *Rationality in Science* (edited by Risto Hilpinen), Reidel, Dordrecht.
1980: Articles on 'Free Logic', 'Indefinite Descriptions' and 'Definite Descriptions' in *Handbuch wissenschaftstheoretischer Begriffe* (edited by Joseph Speck), Vandenhoeck and Ruprecht, Göttingen.
1981: 'On the Philosophical Foundations of Free Logic', *Inquiry* **24**, 147—203.
1981: 'A Speculation on Predication', in *Philosophie Als Wissenschaft* (edited by E. Morscher, *et al.*), Cornes, Bad Reichenall.
1982: 'The Ineliminability of E! in Free Quantification Theory Without Identity' (with R. K. Meyer and E. Bencivenga), *Journal of Philosophical Logic* **11**, 229—231.
1982: 'A Logical Interpretation of Meinong's Principle of Independence', *Topoi* **1**, 87—96.
1983: 'Review of C. J. F. Williams' *What is Existence?*', *Philosophical Books*, pp. 103—108.
1985: 'On the Elimination of Singular Terms', *Logique et Analyse* **108**, 379—392.
1985: 'What is Russell's Theory of Descriptions? An Addendum', *Pacific Philosophical Quarterly* **65**, 140—148.
1986: 'Predication and What There is', in *Mérites et Limites des Méthodes Logiques en Philosophie* (ed. J. Vuillemin), J. Vrin, Paris, pp. 219—225.
1986: 'A Free Logic With Simple and Complex Predicates' (with E. Bencivenga), in *Notre Dame Journal of Formal Logic* **27**, 247—256.
1986: 'Nonexistent Objects: Why Theories About Them are Important', in *Grazer Philosophische Studien* **25—26**, 439—446.
1986: 'Predication and Ontological Commitment', in *The Tasks of Contemporary Philosophy: Proceedings of the 10th International Wittgenstein Symposium*, Holder-Pichler-Tempsky, Wien. Reprinted in *Philosophical Applications of Free Logic* (edited by Karel Lambert), Oxford University Press, New York.
1987: 'Prädikation und Abstraktion', *Zeitschrift für Semiotik* **9**, 97—105.
1987: 'Predication and Ontology', *Canadian Journal of Philosophy* **17**, 603—614.
1987: 'On The Philosophical Foundations of Free Description Theory', *History and Philosophy of Logic* **8**, 57—66.
1988: 'Prolegomenon zu einer Theorie des wissenschaftlichen Verstechens', in *Erklären und Verstehen in der Wissenschaft* (edited by G. Schurz), Oldenbourg Verlag, München, pp. 299—319.
1989: 'Free Logic', forthcoming in *Handbook of Formal Ontology* (edited by B. Smith, *et al.*), Philosophia Verlag, München.

1990: 'Definite Descriptions and the Theory of Objects', forthcoming in *Formal Ontology Today* (edited by K. Mulligan).

1990: 'On whether an answer to a why-question is a scientific explanation if and only if it yields scientific understanding', forthcoming in *Causality, Method, and Modality. Essays for Jules Vuillemin* (edited by G. G. Brittan jr.), Kluwer, Dordrecht, pp. 125–142.

1991: 'A Theory of Definite Descriptions', in *Philosophical Applications of Free Logic* (edited by Karl Lambert), Oxford University Press, New York, pp. 17–27.

1991: 'A Theory About Logical Theories of "Expressions of the Form 'The So and So', Where 'The' is in the Singular" ', forthcoming in *Erkenntnis-Orientated. A Bicentennial Volume for Rudolf Carnap and Hans Reichenbach* (edited by W. Spohn), Kluwer, Dordrecht, pp. 337–346.

INDEX

The University of Western Ontario Series in Philosophy of Science

1. J. Leach, R. Butts and G. Pearce (eds.): *Science, Decision and Value.* 1973
ISBN 90-277-0239-X; Pb 90-277-0327-2
2. C. A. Hooker (ed.): *Contemporary Research in the Foundations and Philosophy of Quantum Theory.* 1973
ISBN 90-277-0271-3; Pb 90-277-0338-8
3. J. Bub: *The Interpretation of Quantum Mechanics.* 1974
ISBN 90-277-0465-1; Pb 90-277-0466-X
4. D. Hockney, W. Harper and B. Freed (eds.): *Contemporary Research in Philosophical Logic and Linguistic Semantics.* 1975
ISBN 90-277-0511-9; Pb 90-277-0512-7
5. C. A. Hooker (ed.): *The Logico-algebraic Approach to Quantum Mechanics.*
Vol. I: Historical Evolution. 1975 ISBN 90-277-0567-4; Pb 90-277-0613-1
Vol. II: Contemporary Consolidation. 1979
ISBN 90-277-0707-3; Pb 90-277-0709-X
6. W. L. Harper and C. A. Hooker (eds.): *Foundations of Probability Theory, Statistical Inference, and Statistical Theories of Science.*
Vol. I: Foundations and Philosophy of Epistemic Applications of Probability Theory. 1976 ISBN 90-277-0616-6; Pb 90-277-0617-4
Vol. II: Foundations and Philosophy of Statistical Inference. 1976
ISBN 90-277-0618-2; Pb 90-277-0619-0
Vol. III: Foundations and Philosophy of Statistical Theories in the Physical Sciences. 1976 ISBN 90-277-0620-4; Pb 90-277-0621-2
7. C. A. Hooker (ed.): *Physical Theory as Logico-operational Structure.* 1979
ISBN 90-277-0711-1
8. J. M. Nicholas (ed.): *Images, Perception, and Knowledge.* 1977
ISBN 90-277-0782-0
9. R. E. Butts and J. Hintikka (eds.): *Logic, Foundations of Mathematics, and Computability Theory.* Part One: Logic, Foundations of Mathematics, and Computability Theory. 1977 ISBN 90-277-0708-1
10. R. E. Butts and J. Hintikka (eds.): *Logic, Foundations of Mathematics, and Computating Theory.* Part Two: Foundational Problems in the Special Sciences. 1977 ISBN 90-277-0710-3
11. R. E. Butts and J. Hintikka (eds.): *Logic, Foundations of Mathematics, and Computability Theory.* Part Three: Basic Problems in Methodology and Linguistics. 1977 ISBN 90-277-0829-0
12. R. E. Butts and J. Hintikka (eds.): *Logic, Foundations of Mathematics, and Computability Theory.* Part Four: Historical and Philosophical Dimensions of Logic, Methodology and Philosophy of Science. 1977 ISBN 90-277-0831-2
Set (9-12) ISBN 90-277-0706-5
13. C. A. Hooker J. J. Leach and E. F. McClennen (eds.): *Foundations and Applications of Decision Theory.*
Vol. I: Theoretical Foundations. 1978 ISBN 90-277-0842-8
Vol. II: Epistemic and Social Applications. 1978 ISBN 90-277-0844-4

The University of Western Ontario Series in Philosophy of Science

14. R. E. Butts and J. C. Pitt (eds.): *New Perspectives on Galileo.* 1978 ISBN 90-277-0859-2; Pb 90-277-0891-6
15. W. L. Harper, R. Stalmaker and G. Pearce (eds.): *Ifs. Conditionals, Belief, Decision, Chance, and Time.* 1981 ISBN 90-277-1184-4; Pb 90-277-1220-4
16. J. C. Pitt (ed.): *Philosophy in Economics.* 1981 ISBN 90-277-1210-7; Pb 90-277-1242-5
17. M. Ruse: *Is Science Sexist?* And Other Problems in the Biomedical Sciences. 1981 ISBN 90-277-1249-2; Pb 90-277-1250-6
18. N. Rescher: *Leibniz's Metaphysics of Nature.* A Group of Essays. 1981 ISBN 90-277-1252-2; Pb 277-1253-0
19. L. Laudan: *Science and Hypothesis.* Historical Essays on Scientific Methodology. 1981 ISBN 90-277-1315-4
20. W. R. Shea (ed.): *Nature Mathematized.* Historical and Philosophical Case Studies in Classical Modern Natural Philosophy. 1983 ISBN 90-277-1402-9
21. M. Ruse (ed.): *Nature Animated.* Historical and Philosophical Case Studies in Greek Medicine, 19th-Century and Recent Biology, Psychiatry, and Psychoanalysis. 1983 ISBN 90-277-1403-7
22. W. R. Shea (ed.): *Otto Hahn and the Rise of Nuclear Physics.* 1983 ISBN 90-277-1584-X
23. H. F. Cohen: *Quantifying Music.* The Science of Music at the First Stage of the Scientific Revolution (1580-1650). 1984 ISBN 90-277-1637-4
24. R. E. Butts: *Kant and the Double Government Methodology.* Supersensibility and Method in Kant's Philosophy of Science. 1984 ISBN 90-277-1760-5; Pb 90-277-2384-2
25. J. R. Brown (ed.): *Scientific Rationality.* The Sociological Turn. 1984 ISBN 90-277-1812-1
26. F. Wilson: *Explanation, Causation and Deduction.* 1985 ISBN 90-277-1856-3
27. J. C. Pitt (ed.): *Change and Progress in Modern Science.* 1985 ISBN 90-277-1898-9
28. H. B. Hollinger and M. J. Zenzen: *The Nature of Irreversibility.* A Study of Its Dynamics and Physical Origins. 1985 ISBN 90-277-2080-0
29. K. Okruhlik and J. R. Brown (eds.): *The Natural Philosophy of Leibniz.* 1985 ISBN 90-277-2145-9
30. G. Oddie: *Likeness to Truth.* 1986 ISBN 90-277-2238-2
31. F. Wilson: *Laws and Other Worlds.* A Human Account of Laws and Counterfactuals. 1986 ISBN 277-2232-3
32. J. Earman: *A Primer on Determinism.* 1986 ISBN 90-277-2240-4; Pb 90-277-2241-2
33. R. E. Butts (ed.): *Kant's Philosophy of Physical Science.* Metaphysische Anfangsgründe der Naturwissenschaft (1786-1986). 1986 ISBN 90-277-2309-5; Pb 90-277-2310-9

The University of Western Ontario Series in Philosophy of Science

34. I. B. MacNeill and G. J. Umphrey (eds.): *Advances in the Statistical Sciences.* Festschrift in Honor of Professor V.M. Joshi's 70th Birthday (6 volumes). Vol. I: Applied Probability, Stochastic Processes, and Sampling Theory. 1987 ISBN 90-277-2393-1
35. I. B. MacNeill and G. J. Umphrey (eds.): *Advances in the Statistical Sciences.* Vol. II: Foundations of Statistical Inference. 1987 ISBN 90-277-2394-X
36. I. B. MacNeill and G. J. Umphrey (eds.): *Advances in the Statistical Sciences.* Vol. III: Time Series and Econometric Modelling. 1987 ISBN 90-277-2395-8
37. I. B. MacNeill and G. J. Umphrey (eds.): *Advances in the Statistical Sciences.* Vol. IV: Stochastic Hydrology. 1987 ISBN 90-277-2396-6
38. I. B. MacNeill and G. J. Umphrey (eds.): *Advances in the Statistical Sciences.* Vol. V: Biostatistics. 1987 ISBN 90-277-2397-4
39. I. B. MacNeill and G. J. Umphrey (eds.): *Advances in the Statistical Sciences.* Vol. VI: Actuarial Science. 1987 ISBN 90-277-2398-2
    Set (Nos. 34-39) ISBN 90-277-2399-0
40. N. Rescher: *Scientific Realism.* A Critical Reappraisal. 1987 ISBN 90-277-2442-3; Pb 90-277-2528-4
41. B. Skyrms and W. L. Harper (eds.): *Causation, Chance, and Credence.* 1988 ISBN 90-277-2633-7
42. W. L. Harper and B. Skyrms (eds.): *Causation in Decision, Belief Change and Statistics.* 1988 ISBN 90-277-2634-5
43. R. S. Woolhouse (ed.): *Metaphysics and Philosophy of Science in the 17th and 18th Centuries.* Essays in Honor of Gerd Buchdahl. 1988 ISBN 90-277-2743-0
44. R. E. Butts and J. R. Brown (eds.): *Constructivision and Science.* Essays in Recent German Philosophy. 1989 ISBN 0-7923-0251-6
45. A. D. Irvine (ed.): *Physicalism in Mathematics.* 1989 ISBN 0-7923-0513-2
46. J. van Cleve and R. E. Frederick (eds.): *The Philosophy of Right and Left.* Incongruent Counterparts and the Nature of Space. 1991 ISBN 0-7923-0844-1
47. F. Wilson: *Empiricism and Darwin's Science.* 1991 ISBN 0-7923-1019-5
48. G. G. Brittan, Jr. (ed.): *Causality, Method and Modality.* Essays in Honor of Jules Vuillemin. 1991 ISBN 0-7923-1045-4
49. W. Spohn, B. C. van Fraassen and B. Skyrms (eds.): *Existence and Explanation.* Essays in Honor of Karel Lambert. 1991 ISBN 0-7923-1252-X

MIX
Papier aus verantwortungsvollen Quellen
Paper from responsible sources
FSC® C105338

If you have any concerns about our products,
you can contact us on
ProductSafety@springernature.com

In case Publisher is established outside the EU,
the EU authorized representative is:
Springer Nature Customer Service Center GmbH
Europaplatz 3, 69115 Heidelberg, Germany

Printed by Libri Plureos GmbH
in Hamburg, Germany